THE BEHAVIORAL, MOLECULAR, PHARMACOLOGICAL, AND CLINICAL BASIS OF THE SLEEP–WAKE CYCLE

THE BEHAVIORAL, MOLECULAR, PHARMACOLOGICAL, AND CLINICAL BASIS OF THE SLEEP–WAKE CYCLE

Edited by

Eric Murillo-Rodríguez
Universidad Anáhuac Mayab, Mérida, Yucatán, Mexico

ELSEVIER

ACADEMIC PRESS
An imprint of Elsevier

Academic Press is an imprint of Elsevier
125 London Wall, London EC2Y 5AS, United Kingdom
525 B Street, Suite 1650, San Diego, CA 92101, United States
50 Hampshire Street, 5th Floor, Cambridge, MA 02139, United States
The Boulevard, Langford Lane, Kidlington, Oxford OX5 1GB, United Kingdom

Notices

Knowledge and best practice in this field are constantly changing. As new research and experience broaden our understanding, changes in research methods, professional practices, or medical treatment may become necessary.

Practitioners and researchers must always rely on their own experience and knowledge in evaluating and using any information, methods, compounds, or experiments described herein. In using such information or methods they should be mindful of their own safety and the safety of others, including parties for whom they have a professional responsibility.

To the fullest extent of the law, neither the Publisher nor the authors, contributors, or editors, assume any liability for any injury and/or damage to persons or property as a matter of products liability, negligence or otherwise, or from any use or operation of any methods, products, instructions, or ideas contained in the material herein.

British Library Cataloguing-in-Publication Data
A catalogue record for this book is available from the British Library

Library of Congress Cataloging-in-Publication Data
A catalog record for this book is available from the Library of Congress

ISBN: 978-0-12-816430-3

Publisher: Nikki Levy
Acquisition Editor: Natalie Farra
Editorial Project Manager: Timothy Bennett
Production Project Manager: Kiruthika Govindaraju
Cover Designer: Greg Harris

Typeset by MPS Limited, Chennai, India

Contents

List of Contributors

Mario Eduardo Acosta-Hernández Biomedicine Department, Health Sciences Institute, Veracruzana University, Veracruz, Mexico

Aljohara S. Almeneessier University Sleep Disorders Center, Department of Medicine, College of Medicine, King Saud University, Riyadh, Saudi Arabia; Department of Family Medicine, College of Medicine, King Saud University, Riyadh, Saudi Arabia

Ahmed S. BaHammam University Sleep Disorders Center, Department of Medicine, College of Medicine, King Saud University, Riyadh, Saudi Arabia

Edward O. Bixler Sleep Research & Treatment Center, Penn State Health Milton S. Hershey Medical Center, Penn State College of Medicine, Hershey, PA, United States

Ritchie E. Brown Psychiatry, VA Boston Healthcare System and Harvard Medical School, West Roxbury, MA, United States

Henning Budde Intercontinental Neuroscience Research Group; Faculty of Human Sciences, Medical School Hamburg, Hamburg, Germany; Physical Activity, Physical Education, Health and Sport Research Centre (PAPESH), Sports Science Department, School of Science and Engineering, Reykjavik University, Reykjavik, Iceland; Lithuanian Sports University, Kaunas, Lithuania

Jacqueline Danisi Sleep Research & Treatment Center, Penn State Health Milton S. Hershey Medical Center, Penn State College of Medicine, Hershey, PA, United States

Subimal Datta Department of Anesthesiology, Graduate School of Medicine, The University of Tennessee, Knoxville, TN, United States; Department of Psychology, College of Arts and Sciences, The University of Tennessee, Knoxville, TN, United States

Julio Fernandez-Mendoza Sleep Research & Treatment Center, Penn State Health Milton S. Hershey Medical Center, Penn State College of Medicine, Hershey, PA, United States

Fabio García-García Biomedicine Department, Health Sciences Institute, Veracruzana University, Veracruz, Mexico

Edgar Garcia-Rill Center for Translational Neuroscience, University of Arkansas for Medical Sciences, Little Rock, AR, United States

Ravi Gupta Department of Psychiatry, All India Institute of Medical Sciences, Rishikesh, India

Sérgio Machado Intercontinental Neuroscience Research Group; Laboratory of Panic and Respiration, Institute of Psychiatry, Federal University of Rio de Janeiro, Rio de Janeiro, Brazil; Salgado de Oliveira University, Rio de Janeiro, Brazil; Physical Activity Neuroscience Laboratory, Physical Activity Sciences Postgraduate Program, Salgado de Oliveira University (UNIVERSO), Rio de Janeiro, Brazil

Diogo Monteiro Intercontinental Neuroscience Research Group; Sport Science School, Polytechnic Institute of Santarém, Santarém, Portugal

Jaime M. Monti Department of Pharmacology and Therapeutics, School of Medicine, Clinics Hospital, Universidad de la República, Montevideo, Uruguay

Eric Murillo-Rodríguez Molecular and Integrative Neuroscience Laboratory, School of Medicine, Health Science Division, Universidad Anahuac Mayab, Mérida, Yucatán, México; Intercontinental Neuroscience Research Group

Michael D. Oliver Department of Anesthesiology, Graduate School of Medicine, The University of Tennessee, Knoxville, TN, United States; Department of Neurology, Graduate School of Medicine, The University of Tennessee, Knoxville, TN, United States

Seithikurippu R. Pandi-Perumal Somnogen Canada Inc., Toronto, ON, Canada

Oscar Ramírez-Plascencia Biomedicine Department, Health Sciences Institute, Veracruzana University, Veracruz, Mexico; Medicine School, Autonomous University of San Luis Potosí, San Luis Potosi, Mexico

Nuno Barbosa Rocha Intercontinental Neuroscience Research Group; Health School, Polytechnic Institute of Porto, Porto, Portugal

Timothy Roehrs Sleep Disorders and Research Center, Henry Ford Health System, Detroit, MI, United States; Department of Psychiatry and Behavioral Neurosciences, School of Medicine, Wayne State University, Detroit, MI, United States

Thomas Roth Sleep Disorders and Research Center, Henry Ford Health System, Detroit, MI, United States; Department of Psychiatry and Behavioral Neurosciences, School of Medicine, Wayne State University, Detroit, MI, United States

Pradeep Sahota Harry S. Truman Memorial Veterans Hospital, Columbia, MO, United States; Department of Neurology, University of Missouri, Columbia, MO, United States

Rishi Sharma Harry S. Truman Memorial Veterans Hospital, Columbia, MO, United States; Department of Neurology, University of Missouri, Columbia, MO, United States

Diogo Telles-Correira Intercontinental Neuroscience Research Group; Faculty of Medicine, University of Lisbon, Lisbon, Portugal

Mahesh M. Thakkar Harry S. Truman Memorial Veterans Hospital, Columbia, MO, United States; Department of Neurology, University of Missouri, Columbia, MO, United States

Pablo Torterolo Laboratory of Sleep Neurobiology, Department of Physiology, School of Medicine, Universidad de la República, Montevideo, Uruguay; Intercontinental Neuroscience Research Group

Giancarlo Vanini Department of Anesthesiology, University of Michigan, Ann Arbor, MI, United States

André Barciela Veras Intercontinental Neuroscience Research Group; Dom Bosco Catholic University, Campo Grande, Mato Grosso, Brazil

Alexandros N. Vgontzas Sleep Research & Treatment Center, Penn State Health Milton S. Hershey Medical Center, Penn State College of Medicine, Hershey, PA, United States

Tetsuya Yamamoto Intercontinental Neuroscience Research Group; Graduate School of Technology, Industrial and Social Sciences, Tokushima University, Tokushima, Japan

Introduction

Although many within our society subscribe to the view that "sleep is for the weak," there is increasing recognition within the scientific and medical community that adequate amounts of high quality, uninterrupted sleep are essential for the health of both the brain and the body.[1] While there is considerable variability between individuals, the consensus statement of the American Academy of Sleep Medicine and the Sleep Research Society advises that adults should obtain at least 7 hours of uninterrupted sleep a night on a consistent basis.[2] However, many of us do not reach this value or have fragmented sleep due to social schedules, work requirements, parental duties, or an underlying medical condition. In fact, the diagnosis of sleep disorders is increasing. For instance, there has been a sixfold increase in the diagnosis of sleep disorders within the US veteran population in the period from 2000 to 2010.[3] Furthermore, disrupted sleep physiology is now recognized as contributing to severe disorders such as schizophrenia,[4] Alzheimer's disease,[5] alcoholism,[6] and the epidemic of obesity.[7] Thus, treatment of sleep disruption, by cognitive behavioral therapy, pharmacological means or noninvasive stimulation, is increasingly being considered as an important component of a comprehensive treatment plan.

Considerable progress has been made in the last hundred years in our understanding of the brain mechanisms that control the sleep–wake cycle,[8] culminating in the award of the 2017 Nobel Prize in Physiology or Medicine to Hall, Rosbash, and Young for their elucidation of the molecular mechanisms that regulate circadian rhythms.[9] Thus, this state-of-the-art book, which focuses on the basic mechanisms underlying the sleep–wake cycle and how they are affected in clinical conditions, is timely. Nine standalone chapters by leading experts in the field review the basic physiology, pharmacology, and genetics of the sleep–wake cycle and circadian rhythms. In Chapter 1, The Sleep–Wake Cycle: An Overview, by Roehrs and Roth provide a concise but comprehensive overview of the clinical assessment of sleep, including classic polysomnography recordings and alterations in autonomic measures. A brief overview of the neural pathways of the ascending reticular activating system (ARAS) and circadian and homeostatic control is provided, as well as factors that may affect sleep such as age, temperature, drugs, and shift work. In Chapter 2, Electrophysiological

Correlates of the Sleep–Wake Cycle, by Oliver and Datta contrast the assessment of sleep in humans and in animal models, primarily cats and rodents, with a focus on the different stages of sleep, which may be differentially impacted by drugs and in medical conditions such as schizophrenia.[4] Animal models allow investigation of the cellular mechanisms that may be involved. In Chapter 3, Physiological Mechanisms for the Control of Waking, by Garcia-Rill focuses on one brain electrical phenomenon, which is a feature of conscious brain states, namely increased activity within the gamma band (30–80 Hz or higher) in the cerebral cortex. He reviews work from his laboratory that has characterized cellular mechanisms within the dorsal arm of the ARAS (pedunculopontine nucleus of the brainstem and parafascicular thalamic nucleus), which may promote gamma-frequency discharge of the neurons in this pathway, which is damaged in patients with disorders of consciousness.[10] Gamma band activity is also controlled by the ventral arm of the ARAS, which passes through the basal forebrain.[11] Thus, control of this activity is an important function of the ARAS. In addition to disorders of consciousness, alterations in gamma band activity have been reported in schizophrenia,[12] Alzheimer's disease,[13] and autism.[14] Thus, understanding the cellular and circuit mechanisms has broad clinical significance. Chapter 4, Neurochemistry and Pharmacology of Sleep, by Torterolo, Monti, and Vanini gives a comprehensive review of the neurotransmitter systems that impact the generation of sleep and wakefulness. They present models of interacting neurotransmitter systems that seek to explain the mechanisms that lead to switching between the states of wakefulness, non–rapid eye movement (NREM) and rapid eye movement (REM) sleep. They also provide a useful summary of clinically used pharmacological agents that act on these neurotransmitter systems. Chapter 5, The Role of Neuroglobin in Brain Function and Sleep–Wake Cycle, by Plasencia, Acosta-Hernandez, and Garcia-Garcia describes a highly conserved but poorly understood protein, neuroglobin, a relative of hemoglobin and myoglobin, which is expressed in the brain within sleep–wake control neurons and may play an important neuroprotective role. Chapter 6, Overview of Sleep Disorders, by BaHamman and colleagues describes the clinical presentation, assessment and treatment of common sleep disorders such as obstructive sleep apnea, hypersomnolence, narcolepsy, and movement-related sleep disorders. Chapter 7, Obesity and Sleep Disturbances, by Danisi, Fernandez-Mendoza, and Vgontzas covers important work that has revealed the bidirectional interactions between sleep disturbances and metabolic disorders. The potential role of proinflammatory cytokines in mediating these interactions is discussed. Chapter 8, Sleep Disorders and Genes, by Murillo-Rodriguez, Torterolo, and colleagues provides a synopsis of recent studies that have linked alterations in particular genes to sleep and circadian rhythm disorders. The intriguing possibility that the cutting edge gene editing technique, CRISPR, may be useful in treating these disorders is explored. Finally, in Chapter 9, Alcoholism and Sleep, by Sharma, Sahota, and Thakkar give a state-of-the-art review of the literature and their recent work in rodent models, which investigates the mechanisms by which alcohol affects sleep and how disruption of sleep may further alcohol use disorder. In particular, a role for alterations of sleep homeostatic mechanisms in the basal forebrain is proposed as a contributing factor.

Collectively, these chapters summarize current knowledge of the brain mechanisms that control sleep and wakefulness and their timing. They provide the basis for understanding sleep disorders and how disrupted sleep–wake mechanisms can contribute to disorders of consciousness, schizophrenia, neurodegenerative disorders, obesity, and alcoholism. As such, this book will be a useful resource for medical students, residents, clinicians, and researchers.

Ritchie E. Brown

Psychiatry, VA Boston Healthcare System and Harvard Medical School, West Roxbury, MA, United States

ACKNOWLEDGMENTS

This work was supported by the US Department of Veterans Affairs Biomedical Laboratory Research and Development Service Merit Award I01BX001356, and by the US National Institutes of Health (NIMH R01 MH039683, NINDS R21 NS093000, NHLBI HL095491). REB is a Research Health Scientist at VA Boston Healthcare System and an Associate Professor of Psychiatry at Harvard Medical School. The contents do not reflect the views of the US Department of Veterans Affairs or the United States Government.

This work was supported by the University of California Institute for Mexico and the United States (UC MEXUS) and Consejo Nacional de Ciencia y Tecnología (CONACyT; Grant # CN-17-19) given to Eric Murillo-Rodriguez.

References

1. Luyster FS, et al. Sleep: a health imperative. *Sleep* 2012;**35**:727–34.
2. Consensus Conference P, et al. Recommended amount of sleep for a healthy adult: a joint consensus statement of the American Academy of Sleep Medicine and Sleep Research Society. *J Clin Sleep Med* 2015;**11**:591–2.
3. Alexander M, et al. The National Veteran Sleep Disorder Study: descriptive epidemiology and secular trends, 2000–2010. *Sleep* 2016;**39**:1399–410.

4. Manoach DS, Pan JQ, Purcell SM, Stickgold R. Reduced sleep spindles in schizophrenia: a treatable endophenotype that links risk genes to impaired cognition? *Biol Psychiatry* 2016;**80**:599–608.
5. Mander BA, Winer JR, Jagust WJ, Walker MP. Sleep: a novel mechanistic pathway, biomarker, and treatment target in the pathology of Alzheimer's disease? *Trends Neurosci* 2016;**39**:552–66.
6. Roehrs T, Roth T. Insomnia as a path to alcoholism: tolerance development and dose escalation. *Sleep* 2018;**41**.
7. Reutrakul S, Van Cauter E. Sleep influences on obesity, insulin resistance, and risk of type 2 diabetes. *Metabolism* 2018;**84**:56–66.
8. Brown RE, Basheer R, McKenna JT, Strecker RE, McCarley RW. Control of sleep and wakefulness. *Physiol Rev* 2012;**92**:1087–187.
9. Sehgal A. Physiology flies with time. *Cell* 2017;**171**:1232–5.
10. Schmahmann JD. Vascular syndromes of the thalamus. *Stroke* 2003;**34**:2264–78.
11. Kim T, et al. Cortically projecting basal forebrain parvalbumin neurons regulate cortical gamma band oscillations. *Proc Natl Acad Sci U S A* 2015;**112**:3535–40.
12. Kwon JS, et al. Gamma frequency-range abnormalities to auditory stimulation in schizophrenia. *Arch Gen Psychiatry* 1999;**56**:1001–5.
13. Palop JJ, Mucke L. Network abnormalities and interneuron dysfunction in Alzheimer disease. *Nat Rev Neurosci* 2016;**17**:777–92.
14. An KM, et al. Altered gamma oscillations during motor control in children with autism spectrum disorder. *J Neurosci* 2018;**38**:7878–86.

CHAPTER

1

The Sleep–Wake Cycle: An Overview

Timothy Roehrs[1,2] *and Thomas Roth*[1,2]

[1]Sleep Disorders and Research Center, Henry Ford Health System, Detroit, MI, United States [2]Department of Psychiatry and Behavioral Neurosciences, School of Medicine, Wayne State University, Detroit, MI, United States

OUTLINE

Molecular Sleep–Wake Cycle
DOI: https://doi.org/10.1016/B978-0-12-816430-3.00001-4

1.1 INTRODUCTION

The sleep–wake cycle in healthy humans is a 24-hour cycle composed of approximately one-third sleep and two-thirds wake. The sleep–wake cycle is under complex, interacting circadian and homeostatic processes. Within the 24-hour sleep–wake cycle is a 90–120 minutes ultradian cycle (basic rest activity cycle), most clearly evident during sleep, but also hypothesized as being present during wakefulness. Sleep and circadian bioscientists continue to amass information regarding the genetic and neurobiological mechanisms underlying the sleep–wake cycle with information about some of the basic features and mechanisms emerging, but there is much yet to be discovered.

Sleep is a vital behavior with the appetitive and essential nature of sleep clearly evident in a human's inability to maintain wakefulness for more than 2 or 3 days. As the state of *sleep need* progressively increases during periods of sustained wakefulness (i.e., sleep deprivation), brief microsleeps begin to intrude into wakefulness during ongoing behavior and during periods of inactivity. As *sleep drive* further increases it is expressed as longer episodes of unintended sleep (i.e., naps).[1] This vital, compulsory nature of sleep is in contrast to one's ability to food or fluid deprive oneself to death.

Sleep is characterized by a stereotypic posture, minimal movement, reduced responsivity to stimuli, reversibility, and species-specific diurnal timing and duration. In humans, sleep is recognized behaviorally by recumbence and eye closure, but some mammals sleep with eyes open (e.g., cattle) or while standing (e.g., horse, elephant).[2] The immobility of human sleep is relative in that sleep walking and talking occur in some human sleep disorders and among animals some fish swim in place and mammals move periodically. The sleep state can be differentiated from death, coma, and hibernation by the characteristics of arousability and rapid reversibility. Sensory (nonvisual) monitoring of both exogenous and endogenous stimulation continues during sleep such that, for example, the vital stimulus of hypoxemia arouses even a severely sleep deprived individual and parents arouse to the cry of their baby. Further, sensory discrimination occurs as the parent does *not* arouse to the cry of another baby whose cry is of a similar stimulus intensity. Among mammals the daily duration of sleep varies from 2 to 20 hours with that of humans being approximately 8 hours.[3] Larger animals have less daily sleep; for example, elephants sleep about 3 hours per day, while the chipmunk sleeps about 16 hours. Sleep is very light or absent during migration or postpartum in some birds and fish and northern fur seals sleep with one half of the brain at a time. Sleep in adult humans, in many, but not all cultures, occurs as a single bout during the dark hours, while for various other mammals sleep occurs in multiple bouts and for some mammals sleep is linked to the light period.

Sleep scientists measure sleep electrophysiologically, as behavioral assessment of sleep and its intensity by testing arousability or reversibility is obtrusive and disruptive of the very state being assessed.[4] Electrophysiological measures correlate well with behavioral observations, but they further reveal subtleties that are not apparent behaviorally and subjectively. For example, some sleep disorders are associated with brief (3–15 seconds) electroencephalographic (EEG) arousals of which the sleeping individual is unaware. The simultaneous recording of the EEG, the electrooculogram (EOG), and the electromyogram (EMG) are the accepted standard measures of sleep and waking and together these measures are termed polysomnography (PSG).[4]

1.2 ELECTROPHYSIOLOGY OF SLEEP: POLYSOMNOGRAPHY

Behind the closed eyes and relative behavioral quiescence of sleep is an active, complex, and highly organized process composed of two distinct brain states: nonrapid eye movement (NREM) sleep and rapid eye movement (REM) sleep. As will be seen below this distinction goes much beyond the presence or absence of eye movements for which these two states are named. Fig. 1.1 depicts PSGs of wake NREM and REM sleep.

We describe the characteristics of a PSG (i.e., EEG, EOG, and EMG) of sleep–wake in detail. In contrast to the low voltage (10–30 μV) and fast frequency (16–35 Hz) of activated wakefulness, the cortical EEG (C3/4-A1/2) of relaxed, eyes-closed wakefulness is characterized by increased voltage (20–40 μV) and an 8–12 Hz frequency. During the transition to sleep, sometimes called drowsy sleep or transitional sleep, the EEG frequency becomes mixed while the voltage remains at the level of relaxed wakefulness. In NREM sleep EEG voltage is further

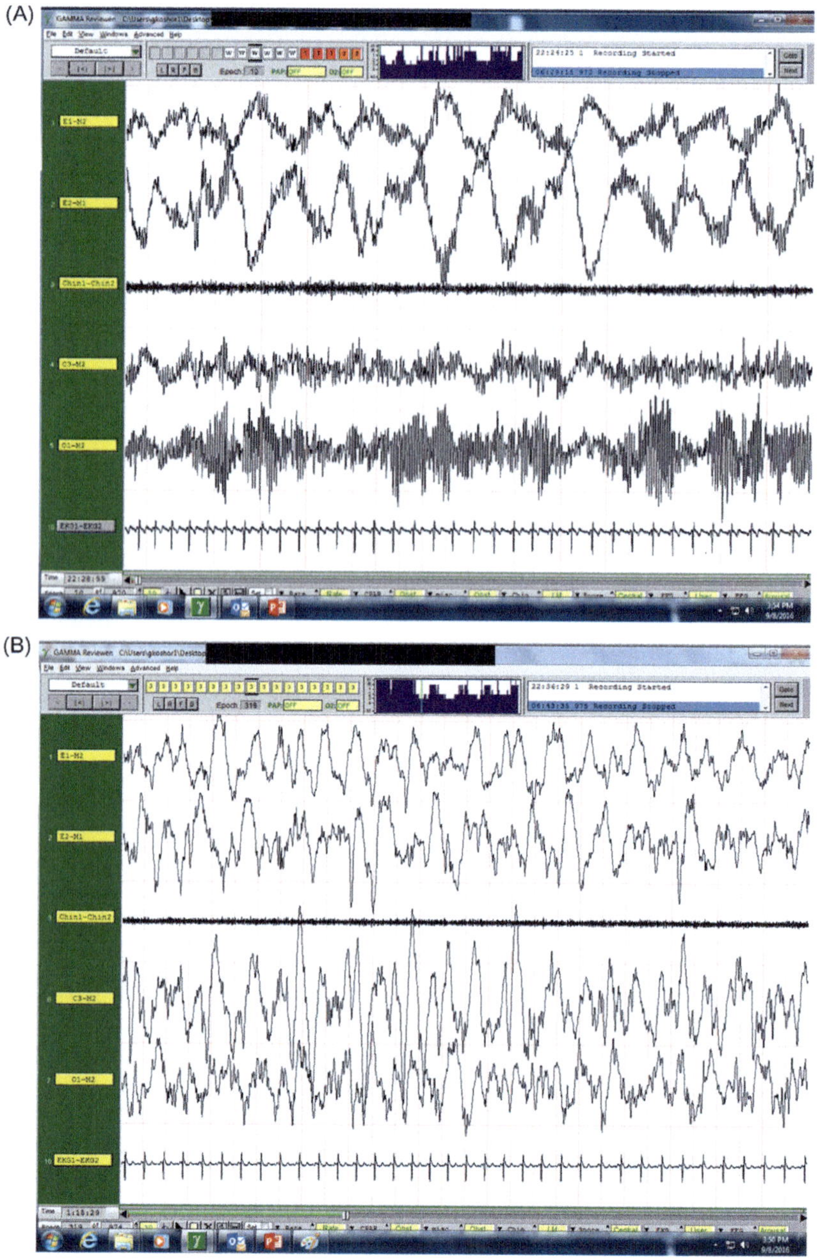

FIGURE 1.1 Displays 30-s screen shots of a standard polysomnogram collected using Grass software. On each panel, channel one and two are left and right EOG, channel three a EMG, channel four a C3-EEG, channel five a 01-EEG, and channel six a EKG. (A) A typical wake epoch, (B) NREM epoch, (C) REM epoch.

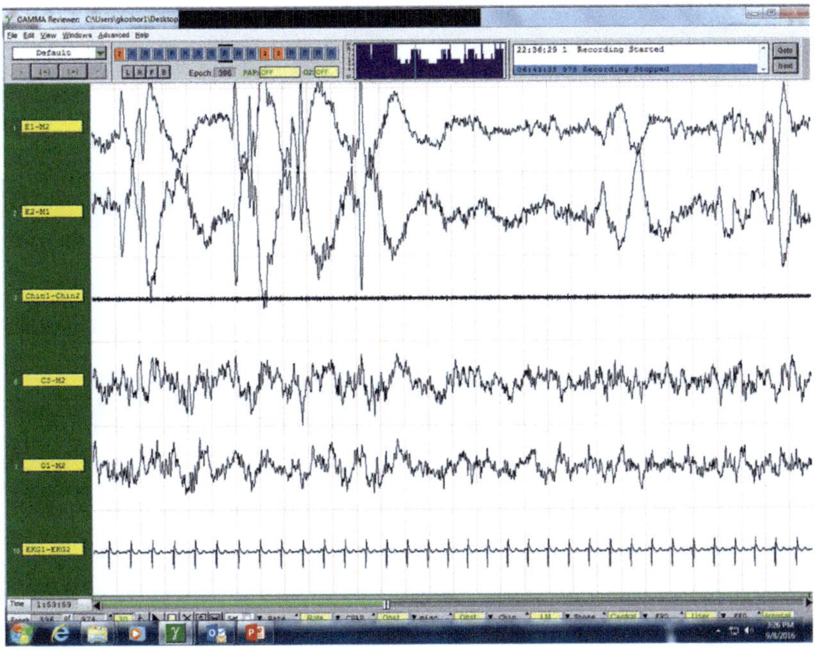

FIGURE 1.1 (Continued).

increased and frequency is further slowed. When arousal threshold is highest, the EEG of NREM sleep has a 0.5–2.5 Hz frequency with voltages of 75 μV and higher, which is termed slow wave sleep (SWS). The EMG, highest in wakefulness, is gradually reduced during NREM sleep, although limb and body movements occur periodically during NREM and there still is voluntary control of musculature.

The EOGs of wakefulness reveal REMs, which during the transition to NREM sleep (i.e., drowsy sleep) become slow and rolling. Importantly, the rolling eye movements mark the onset of the functional blindness that all humans experience during sleep. This contributes to the dangers of drowsy driving. The EOG becomes quiescent during NREM SWS. After the first 90–120 minutes of NREM sleep the healthy normal person enters REM sleep.

The EOG of REM sleep, for which this sleep state is named, is characterized by rapid conjugate eye movements. The cortical EEG of REM reverts to the low voltage, mixed frequency pattern of drowsy sleep. The second defining characteristic of REM sleep is its skeletal muscle atonia, which is reflected in the EMG achieving its lowest level of the 24-hour period. The skeletal muscle atonia of REM sleep occurs through a process of postsynaptic inhibition of motor neurons at the dorsal horn of the spinal cord. Another important feature of REM sleep is its tonic

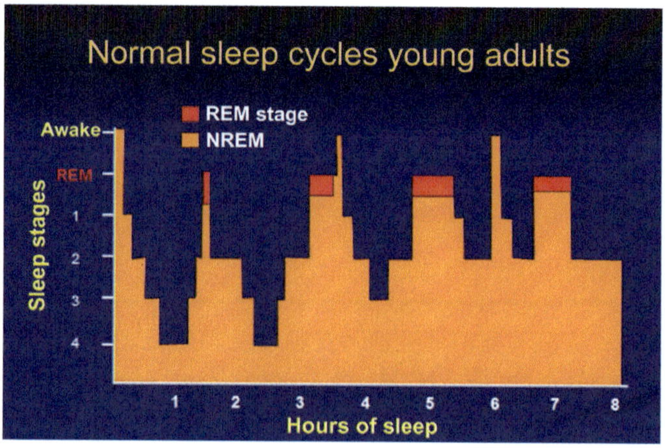

FIGURE 1.2 A histogram that illustrates the NREM-REM cycles across an 8-h sleep period. Source: *Berger RJ. The sleep and dream cycle. In: Kales A, editor.* Sleep physiology and pathology: a symposium. *Philadelphia, PA: Lippincott; 1969.*

and phasic components. The tonic components of REM sleep are the persistent muscle atonia and the desynchronized EEG. The phasic components are intermittent and include bursts of eye movements occurring against a background of EOG quiescence. Coupled with the eye movement bursts are muscle twitches, typically involving peripheral muscles. These twitches are superimposed on the tonic muscle atonia of REM and probably reflect sympathetic drive breaking through the postsynaptic motor inhibition (see Section 1.3.1).

Fig. 1.2 illustrates the progression of sleep stages in a healthy young adult across an 8-hour sleep period. NREM and REM sleep alternate in 90- to 120-minute cycles with the predominance of NREM SWS occurring in the first 4 hours of the night and REM sleep occurring in the last 4 hours.[5] Of note across the night the duration of NREM SWS episodes diminish, while the duration of REM sleep episodes increase. This differential distribution of sleep stages across the night is associated with the unique sleep state–specific physiological and cognitive function changes we will now describe.

1.3 PHYSIOLOGICAL AND COGNITIVE FUNCTION DURING SLEEP

1.3.1 Autonomic Nervous System

The activity of the autonomic nervous system (ANS) varies between the two sleep states (NREM and REM) and the wake state.[6] Parasympathetic activity increases during NREM relative to wakefulness. It remains relatively increased during both tonic and phasic REM. Sympathetic activity remains constant during wake and NREM and is slightly reduced during tonic REM. Consequently, parasympathetic activity predominates during sleep with the exception of phasic REM. Sympathetic drive is dramatically increased during phasic REM and it predominates despite the increased parasympathetic activity of phasic REM.

1.3.2 Respiratory System

Breathing patterns and the control of respiration are altered by sleep state relative to that of waking.[7] Minute ventilation is decreased from waking levels by 13%–15% during NREM sleep. Two factors are responsible; the first being that waking nonmetabolic drive to breathe is removed with the onset of NREM sleep. Secondly, airflow resistance is enhanced due to a reduction of upper airway dilator muscle tone that occurs in conjunction with the general reduction of skeletal muscle tone of sleep. During the tonic skeletal muscle atonia of REM sleep, airway resistance is further increased compared with NREM, resulting in a twofold increase relative to that of waking. This heightened airway resistance coupled with the ANS sympathetic activation, particularly in phasic REM, leads to irregular breathing patterns and even respiratory pauses during REM sleep.

Metabolic control of breathing is also altered by the NREM and REM sleep states.[8] Hypoxic ventilatory drive is reduced in NREM and declines further in phasic REM. Hypercapnic drive, while also reduced in NREM relative to wake, is virtually absent in phasic REM sleep. Breathing during NREM sleep is primarily controlled by arterial levels of CO_2 and thus when levels of CO_2 are below the elevated threshold of NREM, the effort to breathe ceases. Consequently, at transitions from wake to sleep breathing often becomes periodic due to this shifting of hypercapnic set point. Individuals with fragmented sleep characterized by frequent wake-sleep transitions often have central apnea events. For example, the elderly have central apneas that result from fragmented sleep, which themselves further fragment sleep. On the other hand, given the absence of hypercapnic drive in REM sleep, obstructive apneas that occur during REM sleep are prolonged relative to NREM apneas.

1.3.3 Thermal Regulation

The NREM and REM sleep states also display altered thermoregulation.[9] Thermal set point is reduced in NREM sleep relative to that of wake. Consequently, sweating and shivering occur at lower temperatures during NREM compared with wakefulness. Again REM sleep is unique.

There is *no* temperature regulation during REM sleep and sweating and shivering cease during REM. If one remained in REM sleep long enough, body temperature would equilibrate to ambient temperature. However, REM episodes are never much longer than 30 minutes and thus, noticeable body temperature fluctuations do not occur during REM sleep.

1.3.4 Cognition

The awake state with an enhanced *sleep drive* (i.e., excessive daytime sleepiness) and the NREM and REM states are associated with alterations of learning and memory, which may be specific to the stage of the learning-memory process and the type of memory.[1,10] Excessive sleepiness in the wake state impairs acquisition of mnemonic material. Sleep loss studies, sedative drug studies, and studies of patients with disorders of excessive sleepiness have all reported memory impairment. Many of these studies show that the degree of memory impairment is consistent with the degree of waking sleepiness.[1]

The impact of daytime sleepiness on learning and memory is most clearly evident in studies assessing memory at the transition from wake to sleep.[1] Normal subjects were presented stimulus words at 1-minute intervals while falling asleep. When awakened 10 minutes after sleep onset, these subjects could not recall those words presented within the immediate 5-minute wake interval before EEG signs of sleep. That finding has since been replicated and extended to include both explicit and implicit memory tasks, although the replication reported a 3- to 4-minute window of amnesia before sleep onset.[11]

It is hypothesized that after acquisition of new information, sleep serves to facilitate memory consolidation without the potential interference of the constant acquisition of new information that is occurring during wakefulness.[10] Two distinct memory systems are described, declarative and nondeclarative memory, with declarative memory being explicit and accessible, while nondeclarative memory is less explicit and accessible. The dual hypothesis of memory consolidation during sleep suggests that declarative memory is supported by NREM/SWS, and nondeclarative memory by REM sleep.[10] This description simplifies an immense and contentious literature that dates to the discovery of REM sleep, and the interested reader is directed to Rasch and Born.[10]

1.4 REGULATION OF SLEEP AND WAKE

1.4.1 Homeostatic

Homeostatic regulation of sleep is driven by the prior amount of wakefulness relative to the amount of sleep over days and the intensity of *sleep drive* is inferred from measurement of EEG slow wave activity during sleep, arousal threshold during sleep, the total amount and continuity of sleep, and speed of falling asleep at night and during the day.[12] Slow wave activity is distinguished from SWS in that slow waves irrespective of sleep stage can be computer identified and counted. During normal nocturnal sleep, without prior privation, the amount of slow wave activity diminishes in each successive NREM-REM cycle across the night (see Fig. 1.2). Studies have shown increasing prior wakefulness by delaying bedtime or reducing time-in-bed over consecutive nights produces increases in these indices reflecting accumulation of *sleep drive.* The effect is acute as added deprivation or restriction does not lead to continuous enhancement of slow wave activity. Conversely, as expected of homeostatic regulation, reducing *sleep drive* by increasing nocturnal sleep time or daytime napping improves these indices.[13]

Speed of falling asleep during the day also reflects the existence of an underlying homeostatic sleep process. Systematic PSG measurement of daytime sleep latency with a test termed the Multiple Sleep Latency Test (MSLT) has been developed and validated.[1] In healthy volunteers one night of reduced sleep time from 2 to 8 hours by reducing bedtime produces a linear increase in the speed of falling asleep the following day on a standard MSLT. Over consecutive nights of 1–2 hours reduced bedtime, average sleep latency on the MSLT is reduced. Conversely, extension of bedtime beyond 8 hours, or a compensatory nap, increases average sleep latency.[13]

1.4.2 Circadian

Independent of the homeostatic process is a circadian process that organizes sleep and wake and other biological processes (Section 1.4.3) according to the 24-hour daily light-dark cycle.[12] This is a complex system that controls function throughout the body, including sleep and wake. Our understanding of the circadian system is more advanced than that of the homeostatic system. The 2017 Nobel Prize in Physiology or Medicine was awarded to Jeffrey C. Hall, Michael Rosbash, and Michael W. Young "for their discoveries of molecular mechanisms controlling the circadian rhythm."[14] At a cellular level they described genes and proteins that control circadian rhythms. Beyond the cellular level, light-dark input from the retino-hypothalamic tract feeds to the suprachiasmatic nucleus (SCN), which is considered the master biological clock. Efferents from the SCN then convey circadian timing signals that synchronize a variety of physiological systems and organs.

Circadian phase in humans is typically documented by recording body temperature or dim-light onset of melatonin secretion and in rats and mice by wheel running.[14] The nadir in daily human body temperature occurs between 3 and 5 a.m. with a secondary trough occurring over the midday and the peak occurring between 5 and 8 p.m. The duration of sleep episodes and the latency to sleep onset parallels the body temperature rhythm. At temperature peaks latency is delayed and duration is shortened, while at nadirs latency is rapid and duration is enhanced.

The SCN also drives hormonal and metabolic rhythms. To name several, thyroid-stimulating hormone, cortisol, prolactin, growth hormone, and melatonin all show a circadian rhythm. Acutely, prolactin and growth hormone are linked to sleep, meaning their release is delayed when sleep is delayed, regardless of circadian phase. In contrast, cortisol is directly linked to the light-dark cycle and its circadian rhythm remains regardless of the timing of sleep. The hormone melatonin communicates the light-dark cycle through the SCN to clocks throughout the body. Its production and release is controlled by the SCN and it is expressed during darkness and suppressed during light. A short pulse of light during the dark phase suppresses melatonin levels, which continues for the duration of the light. Release of melatonin in the evening at dusk attenuates the alerting pulse of the SCN, thereby facilitating nighttime sleep onset. However, its hypnotic capacity beyond its chronobiotic characteristic, that is, its signaling of darkness, is not clear.

1.4.3 Circadian and Homeostatic Interaction

The most widely accepted model of the interaction of the homeostatic and circadian processes is that of Borbely.[14] In the Borbely two-process model the homeostatic process (process S) builds during wakefulness and decreases during sleep. The circadian process (process C) promotes wake and gates expression of *sleep drive* at the appropriate circadian phase, given that process S has reached its threshold. The biological and molecular substrates of process S, the *sleep drive*, and the neurological pathways by which process C interacts with sleep mechanisms have been partially identified; this is discussed below.

Finally the ultradian rhythm is a 90- to 120-minute cycle of NREM and REM sleep as illustrated in Fig. 1.2. The cycle is repeated three to six times during the night. The homeostatic and ultradian processes appear to be interdependent in that during successive NREM-REM cycles, the amount of SWS per episode declines and the amount of REM sleep per episode increases. The neurobiology underlying this rhythm is only now emerging.

1.5 NEUROBIOLOGICAL CONTROLS OF WAKE AND SLEEP

1.5.1 Wake

In 1949 Moruzzi and Magoun identified an ascending arousal system, which they termed the reticular activation system, and hypothesized that it regulates the level of forebrain wakefulness. Currently, the system is understood as consisting of two main pathways, one ascending to the thalamus and the other extending into the hypothalamus.[15] The thalamic path originates from the cholinergic pedunculopontine and laterodosal tegmental nuclei (PPT-LDT) and the hypothalamic path includes noradrenergic locus coeruleus (LC) and serotonergic dorsal median raphe (DRN) nuclei projections. These projections, joined at the hypothalamus by histaminergic tuberomammillary nucleus (TMN) projections, all together project to the cortex. Electrophysiological recordings from single neurons show LC, DRN, and TMN neurons all fire at their fastest rates during wake, slow during NREM, and are silent in REM. In contrast, PPT-LDT nuclei fire fastest during REM and while also firing during wake, are silent during SWS.

In 1998 two groups independently described pairs of excitatory hypothalamic neuropeptides, hypocretin 1 and 2 and orexin A and B, which later were shown to have a common identity.[16] These neuropeptides are excitatory in nature, directly stimulating LC, DRN, TMN, and the cholinergic basal forebrain (BF) and dopaminergic ventral tegmental area. Maintenance of wakefulness is thought to be stabilized by the activation of the wake systems through an exclusive population of hypocretin/orexin neurons in the posterior lateral hypothalamus (PLH).[16] This hypothesis is supported by both gene knockout studies in mice and human narcolepsy studies showing that the absence of hypocretin/orexin signaling via the type 2 receptor leads to intrusions of sleep into wakefulness.

1.5.2 Sleep

The sleep homeostat, process S, is hypothesized to be mediated by extracellular adenosine (AD).[16] AD is the final breakdown product of ATP, which is involved in intracellular energy turnover. Its levels accumulate with increasing time awake and decline during sleep. Injections of AD into the BF and various other cortical and brainstem targets suggest AD is involved in sleep and wake regulation. Its BF source and anatomical connections to the ventrolateral preoptic nucleus (VLPO), considered the primary locus for initiation of sleep, are unclear.

Sleep initiation occurs through inhibitory pulses generated by neurons in the VLPO, which project to the TMN, LC, and to the PPT-LDT. Most of these projections are GABAergic and galaninergic, which inhibit monoaminergic and cholinergic arousal systems. VLPO neurons fire at their fastest rate during sleep and are silent during wake as shown by electrophysiological recordings. The relation of the VLPO is reciprocal to the monoaminergic and cholinergic arousal systems and the VLPO receives input from these systems, which during wakefulness inhibits the VLPO's sleep-promoting effect.

The VLPO also receives input from the retino-hypothalamic tract through the SCN, which provides circadian signaling for sleep initiation.[15] However, this SCN input is indirect as it projects to the subparaventricular zone, which in turn projects to the dorsomedial hypothalamus (DMH), and finally from the DMH through GABAergic projections to the VLPO.

1.6 FACTORS INFLUENCING SLEEP

1.6.1 Age

Age probably has the greatest influence on sleep, affecting its continuity, staging, and to a lesser degree duration.[5] The sleep of infants is notable for its long duration, with neonates sleeping 16–18 hours daily, which drops to about 14 hours by years 1–2, and 12 hours by year 3. As to staging, REM sleep occupies approximately 50% of sleep time during the first year of life, dropping to about 20% by age 3, where it remains for most of the life span. Slow wave activity during NREM sleep is abundant in the preadolescent years, occupying 20%–30% of sleep time. At adolescence slow wave activity begins a progressive decline that continues throughout adulthood. The decline in slow wave activity has an earlier onset in men compared with women.

In elderly, sleep is characterized by the loss of almost all slow wave activity and sleep duration is reduced to 6–7 hours nightly. Importantly, sleep continuity is disturbed owing to an increase in the prevalence of primary sleep disorders such as sleep apnea/hypopnea and periodic leg movements and to chronic pain and other medical disorders. A possible reduction in *sleep need* and the presence of these other sleep-disruptive factors contribute to the reduced duration and continuity of sleep in elderly. That is both *sleep need* and *sleep ability* are reduced as one ages.

1.6.2 Environment

New sleep environment. An unfamiliar sleep environment is the most universal influence on sleep.[6] A "first-night effect" appearing on the first night sleeping in the sleep laboratory was described early in the modern era of sleep research. Compared with sleep on subsequent nights the first night in the laboratory is characterized by increased sleep latency, increased wakefulness after sleep onset, greater amounts of light stage 1 NREM sleep, and an increased latency to REM sleep. Amounts of slow wave and REM sleep can also be reduced. As a result, PSG research studies of sleep, but not clinical diagnostic studies, will always include an adaptation night.

Noise. Noise appears to be an obvious disrupter of sleep and numerous studies have documented the effects of noise.[6] However, there are wide variations among individuals in their sensitivities and adaptation to noise and the meaning and relevance of the noise is a critical factor. Some bed partners are able to tolerate the most raucous snoring and snorting, while others find soft wheezing offensive and disruptive of their sleep.

Studies of the effects of noise clearly show sleep disturbance, characterized by body movements, awakenings, and sleep stage changes. Light NREM stage 1 and wakefulness are increased, whereas the amount of SWS and REM sleep is reduced. Adaptation generally occurs to the disruptive effects of noise in that complete awakening with recall as *sleep drive* increases is reduced, but brief cortical EEG arousal and EKG reactivity remain.

Temperature. Temperature is another of those obvious disrupters of sleep. Often people with insomnia attribute their sleep difficulty to inadequate room temperature regulation.[6] As with noise there is wide variation in sensitivity and adaptability to room temperature with most able to establish their optimal sleeping temperature. Systematic laboratory studies show people sleeping at temperatures above and below thermos-neutrality exhibit sleep disturbance. Whether cold or hot rooms are more or less disruptive of sleep is not clear, which may relate to differences and preferences among individuals.

1.6.3 Drug Use

Any drug that crosses the blood–brain barrier has a potential effect, dependent on its pharmacokinetic properties, on sleep with the effects being an alteration of the speed of falling asleep or the degree of maintaining sleep.[17] The effects may be immediate and/or are more evident on the discontinuation of use. As might be expected, stimulants such as caffeine, amphetamines, and cocaine delay sleep onset and disrupt sleep, whereas depressants, including alcohol, barbiturates, and benzodiazepine agonists, including both classical benzodiazepines and the "z" drugs, promote sleep onset and enhance sleep continuity. In the case of alcohol this sleep-promoting effect is short-lived. As to sleep staging, alcohol and barbiturates increase SWS, the benzodiazepines suppress it, and the "z" drugs do not alter SWS. Alcohol, barbiturates, opiates, and amphetamines suppress REM sleep, although tolerance to the REM-suppressing effects develops rapidly. In contrast, with a few notable exceptions, most antidepressants suppress REM sleep and tolerance does not develop.

On discontinuation, rebound effects can be observed that are dependent on dose and, possibly, duration of use. As an example, after REM-suppressing drugs are discontinued, enhanced amounts of REM sleep are observed, often with REM sleep fragmented and disrupted by awakenings. Termed REM rebound, these REM sleep disruptions are observed with stimulants, depressants including alcohol, opiates, and antidepressants. Sleep continuity is also disturbed with increased wakefulness during sleep, particularly if tolerance to the drug's sedative effects has developed, as it does with alcohol or barbiturates. All benzodiazepine agonists, at high doses beyond the asymptote of hypnotic effects, will produce sleep disturbance on discontinuation, a phenomenon termed rebound insomnia. With stimulants during discontinuation the opposite effects occur, and enhanced nocturnal sleep and increased daytime sleepiness are seen. The existence of a discontinuance-associated sleep disruption or enhanced daytime sleepiness is quite likely a function of dose and duration of previous administration, although studies have not systematically shown the specifics of the relation.

1.6.4 Shifting Sleep Schedules

In situ shifts of sleep schedule as occurs in shift work or in what is termed "social jet lag" are disruptive of the sleep and alertness of normal sleepers. In situ shifts differ from actual jet lag in that the light-dark cycle does not promote a circadian adaptation to the shifted sleep–wake schedule. The extent and characteristics of the sleep disturbance depend on the direction of the shift, advancing versus delaying the sleep period, the number of hours of schedule misplacement, and the number of days of the shift. Because the light-dark cycle continues to oppose the shifted sleep period, adaptation is never complete.

1.7 SUMMARY

We are now well within the seventh decade of the modern era of sleep research and much is now understood regarding circadian control of the sleep–wake cycle at conceptual, cellular, and neurobiological levels. Less is known regarding homeostatic control of sleep; the neurobiology underlying the wake-dependent increase in *sleep drive* is still unclear. Over the last decade much insight into the neurobiological control of sleep and wakefulness has emerged. Control of sleep and wakefulness involves cholinergic and monoaminergic substrates that promote arousal and GABAergic and galaninergic neurons that promote sleep, with these systems interacting in a reciprocally inhibitory fashion to maintain sleep and wakefulness. Wakefulness is stabilized by activation of the wake promoting cholinergic and monoaminergic systems through an exclusive population of hypocretin/orexin neurons in the PLH.

References

1. Roehrs T, Carskadon MA, Dement WC, Roth T. Daytime sleepiness and alertness. In: Kryger M, Roth T, Dement WC, editors. *Principles and practice of sleep medicine*. 6th ed. Philadelphia, PA: Elsevier; 2016. p. 39–48.
2. Tobler I. Is sleep fundamentally different between mammalian species? *Behav Brain Res* 1995;**69**:35–54.
3. Siegel JM. Sleep in animals: a state of adaptive inactivity. In: Kryger M, Roth T, Dement WC, editors. *Principles and practice of sleep medicine*. 6th ed. Philadelphia, PA: Elsevier; 2016. p. 103–14.
4. Rechtschaffen A, Kales A. *A manual of standardized techniques and scoring system for sleep stages of human sleep*. Los Angeles, CA: Brain Information Service/Brain Research Institute, University of California at Los Angeles; 1968.
5. Carskadon MA, Dement WC. Normal human sleep: an overview. In: Kryger M, Roth T, Dement WC, editors. *Principles and practice of sleep medicine*. 6th ed. Philadelphia, PA: Elsevier; 2016. p. 15–24.
6. Roth T, Roehrs T. An overview of normal sleep and sleep disorders. *Eur J Neurol.* 2000;**7**(S4):3–8.
7. Isa FG, Suratt PM, Remmers JE, editors. *Sleep and breathing*. New York: John Wiley & Sons; 1990.
8. Horner RL, Malhotra A. Control of breathing and upper airways during sleep. In: Broaddus VC, Mason RJ, Ernst JD, editors. *Murray & Nadel's textbook of respiratory medicine*. Philadelphia, PA: Elsevier; 2015. p. 1511–26.
9. Krauchi K, Deboer T. The interrelationship between sleep regulation and thermoregulation. *Front Biosci* 2010;**15**:604–25.
10. Rasch B, Born J. About sleep's role in memory. *Physiol Rev* 2013;**93**:681–766.
11. Wyatt JK, Bootzin RR, Anthony J, Bazant S. Sleep onset is associated with retrograde and anterograde amnesia. *Sleep* 1994;**7**:502–11.
12. Achermann P, Borbely AA. Sleep homeostasis and models of sleep regulation. In: Kryger M, Roth T, Dement WC, editors. *Principles and practice of sleep medicine*. 6th ed. Philadelphia, PA: Elsevier; 2016. p. 377–87.
13. Roehrs T, Shore E, Papineau K, Rosenthal L, Roth T. A two-week sleep extension in sleepy normal. *Sleep* 1996;**19**:576–82.
14. *The Nobel Prize in Physiology or Medicine 2017*. Nobelprize.org. Nobel Media AB 2014; Web June 7, 2018. <http://www.nobelprize.org/nobel_prizes/medicine/laureates/2017/>.

15. Fuller PM, Gooly JJ, Saper CB. Neurobiology of the sleep-wake cycle: sleep architecture, circadian regulation, and regulatory feedback. *J Biol Rhythms* 2006;**21**:482–93.
16. Schwartz MD, Kilduff TS. The neurobiology of sleep and wakefulness. *Psychiatr Clin North Am* 2015;**38**:615–44.
17. Roehrs T, Roth T. Drug-related sleep stage changes. *Sleep Med Clin* 2010;**5**:559–70.

Further Reading

Lu J, Sherman D, Devor M, Saper CB. A putative flip-flop switch for control of REM sleep. *Nature* 2006;**441**:589–94.

CHAPTER

2

Electrophysiological Correlates of the Sleep–Wake Cycle

Michael D. Oliver[1,2] *and Subimal Datta*[1,3]

[1]Department of Anesthesiology, Graduate School of Medicine, The University of Tennessee, Knoxville, TN, United States [2]Department of Neurology, Graduate School of Medicine, The University of Tennessee, Knoxville, TN, United States [3]Department of Psychology, College of Arts and Sciences, The University of Tennessee, Knoxville, TN, United States

OUTLINE

2.1 OVERVIEW OF SLEEP AND THE SLEEP–WAKE CYCLE

Sleep is crucial to individual and societal health and wellness and is one of the most impactful facets of life.[1–5] For example, the National Sleep Foundation recommends that adults receive an average of 8 hours

Molecular Sleep–Wake Cycle
DOI: https://doi.org/10.1016/B978-0-12-816430-3.00002-6

of sleep daily,[6] which in a 24-hour period, reflects one-third of each day. Therefore, if we obtain the recommended daily sleep duration, we as humans will spend an estimated one-third of our lives asleep. One can only imagine how important sleep must be since our bodies spend so much time in this state. But just how important is sleep? Centuries of research show that sleep proves to be vital to human existence.[7–15] Sleep serves as a rejuvenation time in which the majority of the body's internal systems are in an anabolic, or rebuilding, state.[16,17] During this state, the production of proteins increases allowing for cellular growth and repair in systems in the body critical for proper functioning (i.e., immune, muscular, nervous, and skeletal). Interestingly, during this time our body is also removing proteins and other metabolites that are known to be involved in neurodegenerative disorders.[18] These excretory metabolites are then removed from the body during the subsequent wake period. Moreover, deficits in sleep are extremely harmful to the immune[11,19] and endocrine systems[20,21] and have been linked with deficits in memory,[8,9,22–24] cognitive impairment,[25] mood disorders,[26,27] and decreased physical performance,[28,29] as well as increased susceptibility to illness such as diabetes,[30] hypertension,[31] obesity,[32,33] and psychiatric and neurodegenerative disorders.[34–39] Therefore, sleep proves to be significant in health promotion, disease prevention, and basic quality of life.

But how does sleep work? How does our body know when to sleep and when to be awake? These questions can be explained by the body's biological clock and homeostatic regulation of sleep. The sleep–wake cycle is controlled by the interaction of two internal biological mechanisms consisting of sleep–wake homeostasis (i.e., the process by which the body maintains a steady state of internal physiological conditions by generating a drive for the obtainment of sleep, or homeostatic sleep drive) and circadian rhythm (i.e., cyclical changes in the body's internal processes occurring over a 24-hour period).[18,40] Typically in humans, this cycle promotes 8 hours of nocturnal sleep and 16 hours of daytime wakefulness. During this wake period, homeostatic drive for sleep slowly accumulates to initiate the subsequent sleep period. The light-dark cycle, differentiating between day and night, has been shown to be the strongest synchronizing agent of the human biological clock. Thus, quality of sleep and feelings of restfulness are at an optimal level when the human sleep schedule is synchronized with both the internal circadian rhythm, regulated by a group of neurons in the hypothalamus called the suprachiasmatic nucleus,[41,42] and the light-dark cycle is regulated by the earth's tilt on its axis in congruence with the sun. In addition, both quality and quantity of sleep have implications that are reflective of health status. Higher quantity and greater quality is conducive to health, whereas a deficit in one or both adversely affects health.[43–46] As such, there is a bidirectional relationship between these characteristics of sleep and the aforementioned body's internal clock regulating the transition between the periods of sleep and wakefulness. The ability of sleep-inducing GABAergic neurons to inhibit wake-promoting neuronal activity during sleep is reflective of better quality and higher sleep quantity.[47]

Human sleep is predominately mono-phasic, meaning that sleep is consolidated over one period, or phase, during a 24-hour time cycle. The physiological demand of the body dictates the initiation of sleep as the homeostatic drive for sleep increases during wakefulness. In addition, during times in which sleep is lost, accumulation of sleep debt occurs. Sleep debt can be defined as the deficit accrued from not getting enough sleep due to sleep deprivation.[48] This debt must be repaid to the body, and if not, other mental and physical facets such as performance, thinking, memory, cognition, and mood are adversely affected as the body reallocates resources to repay the sleep debt. Examples of demand that allow the body to alter sleep patterns such as sleeping during the dark cycle or achieving the recommended 8 hours of sleep include night shift workers, jet lag, staying up late to complete a task, etc. Alternatively, nonhuman animals sleep multiple times in a 24-hour period and are considered polyphasic. Sleep in nonhuman animals is dependent upon engagement in activities that require wakefulness.[18] Predator–prey relationships play a huge role in the amount and period of sleep in nonhuman animals.[18,49,50] For example, rats are nocturnal creatures and roam predominantly at night to avoid predators in the daytime.

2.2 STAGES OF SLEEP

2.2.1 Human Sleep Architecture

Human sleep can be broken down in two types: rapid eye movement (REM) and nonrapid eye movement (NREM). Normal human sleep architecture consists of adults needing about 7–9 hours of sleep each night with about 80% spent in NREM and 20% of sleep spent in REM. REM sleep occurs about 90 minutes after transitioning from wakefulness to sleep and ideally reoccurs every 90–120 minutes as REM and NREM sleep alternate each night on average roughly 4–6 times per night.[51] There are increasingly longer and deeper periods of REM as the night progresses. This stage derives its name from the marked observable appearance of REMs. Additionally, intense brain activity, rapid breathing, and increases in heart rate and blood pressure occur at levels reflective of a wakeful state. Limbs are temporarily paralyzed, which proves to be a protective factor against acting out dreams. This stage is also the stage in which dreams are the most vivid due to increased brain activity in the limbic system.

In NREM, there is a reduction in physiological activity, brain waves are slower and have greater amplitudes, and autonomic function/balance is better (i.e., decreased breathing, heart rate, and blood pressure). NREM was previously thought to consist of four stages, but has been updated by the American Academy of Sleep Medicine (AASM) in 2007 to reflect only three stages, with the previously outlined stages 3 and 4 combined into one stage known as slow-wave sleep (SWS). Stage 1 NREM sleep refers to the initial transition from wakefulness to sleep. This stage can last several minutes. During this stage, alpha waves in the brain, reflective of alertness, begin to disappear, while theta waves, associated with sleep and quiet wakefulness, begin to appear. Additionally, muscles in the body start to relax and oftentimes hypnic jerks, or sleep twitches, occur while eye movement slows as well. Stage 2 NREM sleep is the stage in which humans spend the majority (about 50%) of their time while asleep. This stage is marked by light sleep before reaching a stage of deep sleep. Eye movements stop, heart rate and breathing decrease, muscles relax progressively more, brain wave activity slows, and sleep spindles form in this stage. The final stage in NREM, or SWS, is where delta brain waves are more prominent. These waves are the slowest waveform and represent a very relaxed state and very low brain activity, often reflective of deep sleep. Therefore, this stage of sleep is aptly referred to as SWS. During this stage, breathing and heart rate levels are at their lowest, muscles are relaxed, and the frequency of brain waves are reduced even further, making this state the most difficult stage from which to be awakened.

2.2.2 Animal Sleep Architecture

Similarly, animals experience seemingly regulated sleep behaviors indicative of an internal circadian biological clock.[40] Moreover, although most animals do exhibit a diurnal sleep pattern, nocturnal animals and diurnal animals have comparable mechanisms of action controlling circadian rhythm. However, as previously mentioned, the predator–prey relationship really dictates animal sleep. For example, animals tending to sleep in safer, more concealed locations (i.e., koala bears in the trees or armadillos underground) tend to be the ones that obtain longer hours of sleep, while animals out in the open and exposed to potential predators (i.e., giraffe and sheep) are the ones with little to no sleep as they must remain alert for survival. Although animal sleep regulation may be similar to humans, only warm-blooded animals have been detected to exhibit observable REM and NREM stages that are comparable with that of human sleep. One striking difference is that the proportion of REM sleep varies between animal types. For example, marine animals like dolphins and killer whales receive little to no REM sleep, as they are known as unihemispheric sleepers. These unihemispheric sleepers shut off one hemisphere of the brain while leaving the other hemisphere active to monitor the environment, be alert for predators, and control physiological body processes. Alternatively, smaller animals such as cats and rats have higher rates of brain metabolism, therefore requiring them to attain more sleep. As such, the most common, and preferred, animal models for sleep research include: the mouse, rat, and cat given the amount of sleep, similarities to human sleep stages, and ease of experimentation. In these animals, NREM–REM sleep cycles are much shorter than in human and nonhuman primates and these cyclic NREM–REM sleep epochs in rodents and cats continue throughout sleep during day and night, except when the animal is engaged in activities that require wakefulness. The cyclic organization of sleep varies within and between species. The period lengths of each REM–NREM sleep epoch increase with brain size across species. Within species, the depth and proportion of the NREM sleep phase in each cycle also increases with brain maturation.

2.3 ELECTROPHYSIOLOGICAL CORRELATES

Identifying stages of sleep has important clinical and research implications. However, it is nearly impossible to identify sleep stages by mere observation. Thus, objective explorations into sleep prove to be vital in terms of sleep stage identification. One important methodology that is used most often is examination of simultaneous recordings of electroencephalogram (EEG), electromyogram (EMG), and electrooculogram (EOG) activity in both humans and nonhuman animals.

2.3.1 In Human Sleep

As previously mentioned, stages of human sleep can be divided into categories characterized by physiological activity. For example, NREM sleep is divided into four stages established by progressive EEG synchronization (high-amplitude and low-frequency) and each stage corresponds to an increasing depth of sleep. In the earliest process of NREM sleep, stage I (also known as N1) is characterized by relatively low amplitude ($<50\,\mu V$) theta frequency activity (4–7 cycles per second (Hz) and vertex sharp waves in the EEG. As NREM sleep progresses, stage II (also known as N2) is characterized by slow (<1 Hz) oscillations with distinctive sleep spindles (waxing and waning of 12–14 Hz waves lasting between 0.5 and 1.0 s) and K-complex (a negative sharp wave followed immediately by a slower positive component) waveforms in the EEG. At the deepest stages of NREM sleep (SWS), stages III and IV, only low-frequency wave activity is present. At stage III (also known as N3), sleep is characterized by the addition of high-amplitude (>100 μV) slow waves (1–4 Hz), with no more than 50% of the EEG record occupied by the slow waves. In stage IV (also known as N4), the EEG record is dominated by high-amplitude (150–250 μV) slow waves (1–4 Hz). Throughout this process (between N1 and N4), EEG frequency is decreasing and the amplitude is increasing, muscle tone detected via EMG progressively declines and may be lost altogether in most of the somatic musculature, and slow rolling eye movements of N1 sleep stage first replace rapid saccadic eye movements of waking and then subside, with the eyes finally assuming a divergent upward gaze upon reaching SWS.

Alternatively, in REM sleep, after varying amounts of time (depending upon the size of the animal and its brain), the progressive set of changes observed in the EEG during NREM reverses itself and the EEG eventually resumes the low-amplitude, fast waveform previously seen in waking. Instead of waking, however, behavioral sleep persists. Muscle tone, at first passively decreased during NREM sleep, is now actively inhibited. Stereotyped bursts of saccadic eye movements, called REMs, appear in the EOG and give this state the name REM sleep. This phase of sleep has also been called activated sleep (due to increased EEG activation) and paradoxical sleep (to signal the maintenance of increased behavioral arousal threshold in the presence of the activated brain). Supplemental to these polysomnographic signs, other REM sleep-specific physiological signs are myoclonic twitches in the facial, digital, and even major proximal skeletal muscles; pronounced fluctuations in cardiorespiratory rhythms and core body temperature; and penile erection in males and clitoral engorgement in females. In addition to these physiological signs, occurrence of vivid dreaming is an important mental experience of REM sleep.

2.3.2 In Animal Sleep

Distinctions between stages of NREM sleep in animal models differ slightly from those in humans.[47,52] In the aforementioned animals, NREM sleep is normally subdivided into two stages (SWS-I and SWS-II). SWS-I is identified by the presence of sleep spindles in the cortical EEG. SWS-II is considered deep sleep and is identified by the presence of high-amplitude, low frequency waves in the cortical EEG (delta sleep). Human NREM-II sleep is comparable with the mouse, rat, and cat SWS-I, whereas stages III and IV NREM sleep in humans are comparable with the animal SWS-II. Additionally, physiological signs of REM sleep in humans are comparable with physiological signs of REM sleep in the rat and cat. Two other physiological signs requiring surgical implantations of recording electrodes can also be used to identify REM sleep in the nonhuman primates, rats, and cats. These two signs are (1) theta rhythm in the hippocampal EEG and (2) spiky field potentials in the pons (P-waves), lateral geniculate nucleus, and occipital cortex [called as ponto-geniculo-occipital spikes]. However, another sleep stage, one between SWS-II and REM sleep, is identified mainly in rats and cats and is called the transitional sleep stage (tS-R). During this transitional stage, the appearance of cortical EEG signs is similar to the cortical EEG of SWS-I and hippocampal and pontine EEG signs are similar in appearance to those during REM sleep, but with less intensity.[52] Since most animal sleep and learning studies have not recorded hippocampal and/or pontine EEG, studies have often considered this stage to be SWS-I. Similarly, in sleep studies of healthy humans, hippocampal and/or pontine EEG cannot be recorded to help identify sleep stages. Therefore, based on EEG signs in humans, this transitional sleep stage has been often incorrectly identified as stage II NREM sleep.

2.4 SUMMARY

Sleep is a time in which cellular growth and repair in the body occur, allowing for proper immune, muscular, nervous, and skeletal functioning. Therefore, given that sleep is a vital part of daily living, the brain allocates resources to ensure sleep is achieved, often at the detriment of other facets of daily living that the body deems unnecessary at the time (i.e., memory formation, emotional processing, physical ability). Identification of electrophysiological correlates of sleep proves crucial to both research and clinical practice given that roughly one-third of human existence is spent sleeping. Explorations into cortical EEG activity allow for conclusions to be drawn regarding clinical implications of sleep deficiency and research implications regarding the effects of specific areas of the brain during certain stages of REM and NREM sleep. For example, REM sleep deficiency is associated with episodes of psychosis that manifest in individuals with Parkinson's disease or other forms of dementia. Alternatively, proper REM sleep has been shown to enhance memory formation and consolidation and promote rejuvenation and restfulness. Moreover, identification of behavioral and physiological activity associated with certain brain waves allows researchers to delve deeper into the effects of sleep by properly identifying which specific sleep stage either human or nonhuman animals are subjected to.

References

1. Foley DJ, Monjan AA, Brown SL, Simonsick EM, Wallace RB, Blazer DG. Sleep complaints among elderly persons: an epidemiologic study of three communities. *Sleep* 1995;**18**(6):425–32.
2. Atkinson G, Davenne D. Relationships between sleep, physical activity and human health. *Physiol Behav* 2007;**90**(2–3):229–35.
3. Shahar E, Whitney CW, Redline S, Lee ET, Newman AB, Javier Nieto F, et al. Sleep-disordered breathing and cardiovascular disease: cross-sectional results of the Sleep Heart Health Study. *Am J Respir Crit Care Med* 2001;**163**(1):19–25.
4. Strine TW, Chapman DP. Associations of frequent sleep insufficiency with health-related quality of life and health behaviors. *Sleep Med* 2005;**6**(1):23–7.
5. Taylor DJ, Mallory LJ, Lichstein KL, Durrence HH, Riedel BW, Bush AJ. Comorbidity of chronic insomnia with medical problems. *Sleep* 2007;**30**(2):213–18.
6. Hirshkowitz M, Whiton K, Albert SM, Alessi C, Bruni O, DonCarlos L, et al. National Sleep Foundation's sleep time duration recommendations: methodology and results summary. *Sleep Health* 2015;**1**(1):40–3.
7. Carskadon MA, Dement WC. Cumulative effects of sleep restriction on daytime sleepiness. *Psychophysiology* 1981;**18**(2):107–13.
8. Datta S. Sleep: learning and memory. In: Koob GF, Le Moal M, Thompson RF, editors. *Encyclopedia of behavioral neuroscience*, vol. 3. Oxford: Academic Press; 2010. p. 218–26.
9. Datta S, Oliver MD. Cellular and molecular mechanisms of REM sleep homeostatic drive: a plausible component for behavioral plasticity. *Front Neural Circuits* 2017;**11**:63.
10. Gottlieb DJ, Redline S, Nieto FJ, Baldwin CM, Newman AB, Resnick HE, et al. Association of usual sleep duration with hypertension: the sleep heart health study. *Sleep* 2006;**29**(8):1009–14.
11. Irwin M. Effects of sleep and sleep loss on immunity and cytokines. *Brain Behav Immun* 2002;**16**(5):503–12.
12. Kato M, Phillips BG, Sigurdsson G, Narkiewicz K, Pesek CA, Somers VK. Effects of sleep deprivation on neural circulatory control. *Hypertension* 2000;**35**(5):1173–5.
13. Killgore WD. *Effects of sleep deprivation on cognition. Progress in brain research*, vol. 185. Elsevier; 2010. p. 105–29.
14. Livingstone MS, Hubel DH. Effects of sleep and arousal on the processing of visual information in the cat. *Nature* 1981;**291**(5816):554–61.
15. Pilcher JJ, Huffcutt AI. Effects of sleep deprivation on performance: a meta-analysis. *Sleep* 1996;**19**(4):318–26.
16. Mackiewicz M, Shockley KR, Romer MA, Galante RJ, Zimmerman JE, Naidoo N, et al. Macromolecule biosynthesis: a key function of sleep. *Physiol Genomics* 2007;**31**(3):441–57.
17. Oswald I. *Sleep as a restorative process: human clues. Progress in brain research*, vol. 53. Elsevier; 1980. p. 279–88.
18. Datta S, MacLean RR. Neurobiological mechanisms for the regulation of mammalian sleep–wake behavior: reinterpretation of historical evidence and inclusion of contemporary cellular and molecular evidence. *Neurosci Biobehav Rev* 2007;**31**(5):775–824.
19. Majde JA, Krueger JM. Links between the innate immune system and sleep. *J Allergy Clin Immunol* 2005;**116**(6):1188–98.

20. AlDabal L, BaHammam AS. Metabolic, endocrine, and immune consequences of sleep deprivation. *Open Respir Med J* 2011;**5**:31.
21. Steiger A. Sleep and endocrinology. *J Intern Med Suppl* 2003;**254**(1):13–22.
22. Gais S, Lucas B, Born J. Sleep after learning aids memory recall. *Learn Mem* 2006;**13** (3):259–62.
23. Maquet P. The role of sleep in learning and memory. *Science* 2001;**294**(5544):1048–52.
24. Stickgold R. Sleep-dependent memory consolidation. *Nature* 2005;**437**(7063):1272.
25. Yaffe K, Falvey CM, Hoang T. Connections between sleep and cognition in older adults. *Lancet Neurol* 2014;**13**(10):1017–28.
26. Ford DE, Cooper-Patrick L. Sleep disturbances and mood disorders: an epidemiologic perspective. *Depress Anxiety* 2001;**14**(1):3–6.
27. Armitage R. Sleep and circadian rhythms in mood disorders. *Acta Psychiatr Scand* 2007;**115**(s433):104–15.
28. Fullagar HH, Skorski S, Duffield R, Hammes D, Coutts AJ, Meyer T. Sleep and athletic performance: the effects of sleep loss on exercise performance, and physiological and cognitive responses to exercise. *Sports Med* 2015;**45**(2):161–86.
29. Reilly T, Edwards B. Altered sleep–wake cycles and physical performance in athletes. *Physiol Behav* 2007;**90**(2–3):274–84.
30. Cappuccio FP, D'elia L, Strazzullo P, Miller MA. Quantity and quality of sleep and incidence of type 2 diabetes: a systematic review and meta-analysis. *Diabetes Care* 2010;**33**(2):414–20.
31. Bansil P, Kuklina EV, Merritt RK, Yoon PW. Associations between sleep disorders, sleep duration, quality of sleep, and hypertension: results from the National Health and Nutrition Examination Survey, 2005 to 2008. *J Clin Hypertens* 2011;**13** (10):739–43.
32. Cappuccio FP, Taggart FM, Kandala NB, Currie A, Peile E, Stranges S, et al. Meta-analysis of short sleep duration and obesity in children and adults. *Sleep* 2008;**31** (5):619–26.
33. Wu Y, Zhai L, Zhang D. Sleep duration and obesity among adults: a meta-analysis of prospective studies. *Sleep Med* 2014;**15**(12):1456–62.
34. Gagnon JF, Postuma RB, Mazza S, Doyon J, Montplaisir J. Rapid-eye-movement sleep behaviour disorder and neurodegenerative diseases. *Lancet Neurol* 2006;**5**(5):424–32.
35. Iranzo A, Santamaria J, Tolosa E. The clinical and pathophysiological relevance of REM sleep behavior disorder in neurodegenerative diseases. *Sleep Med Rev* 2009;**13** (6):385–401.
36. Ong AD, Kim S, Young S, Steptoe A. Positive affect and sleep: a systematic review. *Sleep Med Rev* 2017;**35**:21–32.
37. Baglioni C, Spiegelhalder K, Lombardo C, Riemann D. Sleep and emotions: a focus on insomnia. *Sleep Med Rev* 2010;**14**(4):227–38.
38. Postuma RB, Gagnon JF, Vendette M, Fantini ML, Massicotte-Marquez J, Montplaisir J. Quantifying the risk of neurodegenerative disease in idiopathic REM sleep behavior disorder. *Neurology* 2009;**72**(15):1296–300.
39. Wulff K, Gatti S, Wettstein JG, Foster RG. Sleep and circadian rhythm disruption in psychiatric and neurodegenerative disease. *Nat Rev Neurosci* 2010;**11**(8):589.
40. Borbély AA. A two process model of sleep regulation. *Hum Neurobiol* 1982;**1** (3):195–204.
41. Mistlberger RE. Circadian regulation of sleep in mammals: role of the suprachiasmatic nucleus. *Brain Res Rev* 2005;**49**(3):429–54.
42. Moore RY. Suprachiasmatic nucleus in sleep–wake regulation. *Sleep Med* 2007;**8**:27–33.
43. Kamdar BB, Needham DM, Collop NA. Sleep deprivation in critical illness: its role in physical and psychological recovery. *J Intensive Care Med* 2012;**27**(2):97–111.

44. Steptoe A, O'Donnell K, Marmot M, Wardle J. Positive affect, psychological well-being, and good sleep. *J Psychosom Res* 2008;**64**(4):409–15.
45. Van Cauter E, Spiegel K, Tasali E, Leproult R. Metabolic consequences of sleep and sleep loss. *Sleep Med* 2008;**9**:S23–8.
46. Bradley TD, Floras JS. Obstructive sleep apnoea and its cardiovascular consequences. *Lancet* 2009;**373**(9657):82–93.
47. Datta S. Cellular and chemical neuroscience of mammalian sleep. *Sleep Med* 2010;**11**(5):431–40.
48. Van Dongen H, Rogers NL, Dinges DF. Sleep debt: theoretical and empirical issues. *Sleep Biol Rhythms* 2003;**1**(1):5–13.
49. Lima SL, Rattenborg NC, Lesku JA, Amlaner CJ. Sleeping under the risk of predation. *Anim Behav* 2005;**70**(4):723–36.
50. Campbell SS, Tobler I. Animal sleep: a review of sleep duration across phylogeny. *Neurosci Biobehav Rev* 1984;**8**(3):269–300.
51. Carskadon MA, Dement WC. *Normal human sleep: an overview. Principles and practice of sleep medicine.* 4th ed. Philadelphia, PA: WB Saunders; 2005. p. 13–23.
52. Datta S, Hobson JA. The rat as an experimental model for sleep neurophysiology. *Behav Neurosci* 2000;**114**(6):1239.

CHAPTER

3

Physiological Mechanisms for the Control of Waking

Edgar Garcia-Rill

Center for Translational Neuroscience, University of Arkansas for Medical Sciences, Little Rock, AR, United States

OUTLINE

Molecular Sleep–Wake Cycle
DOI: https://doi.org/10.1016/B978-0-12-816430-3.00003-8

3.1 INTRODUCTION

Two of the most important discoveries in sleep–wake control over the last 10 years are the description of electrical coupling in some neurons of the reticular activating system (RAS) and the presence of intrinsic membrane gamma frequency oscillations in three arousal-related nuclei.[1] Coherence, most commonly mediated by electrical coupling through gap junctions, is the process whereby groups of neurons fire in coordination to create a signal that is mirrored precisely by other groups of neurons.[2] We demonstrated the presence of electrical coupling in the pedunculopontine nucleus (PPN),[3] in one of its ascending targets in the intralaminar thalamus (ILT), the parafascicular nucleus (Pf),[4] and in one of its descending targets, the subcoeruleus nucleus dorsalis,[5] a mechanism that allows these groups of neurons to fire coherently. This mechanism has massive clinical implications. For example, the mechanism of action of the stimulant modafinil, which is used for the treatment of narcolepsy and excessive daytime sleepiness, is to increase coupling via gap junctions, promoting coherence.[6] Most neurons in the RAS that are electrically coupled via gap junctions are GABAergic, so that modafinil-induced coupling decreases input resistance, decreases GABA release, thus disinhibiting a host of other systems, which leads to overall higher coherence in the RAS, thalamocortical systems, and elsewhere. On the other hand, blocking gap junctions leads to rapid loss of consciousness, such as when using the fast-acting anesthetics halothane and propofol, both of which efficiently block gap junctions.[7,8] Oleamide and anandamide both promote sleep by blocking gap junctions.[9–11] The antiulcer agent carbenoxolone, and the antimalarial agents quinine and mefloquine, are also efficient gap junction blockers and share pronounced soporific side effects.[12] Thus, drugs that affect gap junctions also affect sleep–wake control by helping synchronize or desynchronize ensemble activity.[13]

The other aspect of large-scale activity that is essential to neural encoding is firing frequency. The other major discovery in the RAS is the presence of intrinsic gamma band activity in the same three RAS nuclei named above.[14–16] As we will see below, gamma band activity has been associated with attentional mechanisms, sensory perception, and memory. Conditions that block gamma band activity such as anesthetics and disease processes, schizophrenia, and Parkinson's disease (PD) all lead to interrupted gamma band activity, with serious consequences on complex brain processes. Electrical coupling and intrinsic oscillations at dedicated frequencies work together to allow oscillations to persist over a wide range of membrane potentials, making their firing frequency independent of membrane potential,[13,17] and thus maintaining their activity. This chapter will stress the role of gamma band activity in the RAS since it is directly related to the control of waking.

3.2 WAKING UP AND STAYING AWAKE

The organizational features of brain activity for generating conscious awareness was long ago proposed to begin through the activation of the RAS by incoming and continuously changing sensory input, as well as by intrinsically generated activity.[18] More recent evidence suggests that, when we awaken, blood flow first increases in the upper brainstem and thalamus, and later increases in the frontal lobes.[19] Thus, awakening entails a rapid reestablishment of consciousness within a few minutes followed by a relatively slow reestablishment of full awareness. Using positron emission tomography, cerebral blood flow upon waking was found to increase soonest in the brainstem and thalamus, suggesting that the reactivation of these regions underlies the reestablishment of basic consciousness. Over the next 15–20 minutes, blood flow increased primarily in anterior cortical regions. These results question ideas that the cortex is solely responsible for achieving conscious awareness.

Moreover, sleep and waking do not appear to be a continuous, but rather a stepwise process. There are slow transitions in the order of minutes between waking and sleep, and in the transition between slow wave sleep (SWS) and rapid eye movement (REM) sleep.[20] In the absence of strong sensory stimuli to induce waking, waking and sleep appear to be "recruited," and may be part of a stepwise process, in which frequencies decrease from gamma to beta to alpha and lower leading to sleep, or increase from delta to theta to alpha and higher, leading to waking.[21] This stepwise process also suggests that the modulation of waking and sleep is progressive but piecemeal, and that it is the coherence at particular frequencies that determines the particular transition state. That is, the new state may be "recruited" to another level rather than gradually formed. Clinically, this may be reflected in varying levels of consciousness from coma (complete absence of wakefulness), to a vegetative state (achieves partial arousal), to a minimally conscious state (partial preservation of consciousness), or to locked-in syndrome (awareness without movement). These may all represent stages of waking based on the degree of damage to gamma band generating processes.[20]

The original description of the RAS proposed that it participates in tonic or continuous arousal.[22] Moreover, lesions of this region were found to eliminate tonic arousal.[23] Thus, both sensory input and intrinsic activity help maintain gamma band activity for prolonged periods and are critically important for such functions as perception, learning, and memory, not to mention consciousness. It has been suggested that consciousness is associated with continuous gamma band activity, as opposed to an interrupted pattern of activity.[24,25] Clinically, the lack of consistent gamma band activity would alter waking and REM sleep drive, decrease attention, disrupt the binding of perceptions and the like, accounting for a number of symptoms in neurological and psychiatric disorders. Dysregulation of this system thus can elicit a host of abnormal sensory and motor responses, both tonic and phasic, that can account for a number of symptoms expressed in schizophrenia, bipolar disorder, posttraumatic stress disorder, PD, Alzheimer's disease, and certain strokes.[26] It should be noted that gamma band activity is present throughout various brain regions including the cortex,[27–30] thalamus,[31,32] hippocampus,[33] basal ganglia,[34] cerebellum,[35] and, of course, the RAS.[26] But how is gamma band activity generated in the RAS?

3.3 RETICULAR ACTIVATING SYSTEM GAMMA ACTIVITY

3.3.1 Two Calcium Channels

The PPN is the only nucleus in the RAS that is active during both states of high frequency (beta/gamma) EEG activity as in waking and REM sleep. Paradoxically, the EEG pattern during these two markedly different states is very similar, but beta/gamma band activity may not be the same during waking as during REM sleep. Classic studies of recordings of PPN neurons during sleep–wake states revealed neurons that were active during waking and REM sleep, referred to as "Wake-REM on" cells; others were active only during REM sleep, referred to as "REM on" cells; and yet others only during waking, referred to as "Wake on" cells, and none were more active during SWS.[36–41] Gamma band activity has been observed in the cortical EEG of the cat in vivo and in PPN neurons when the animal is active;[42] in the region of the PPN in humans during stepping, but not at rest;[43] and firing at low frequencies ~10 Hz at rest in the primate, but firing at gamma band frequencies when the animal woke up, or when the animal began walking on a treadmill.[44] Thus, the same cells were involved in both arousal and motor control in the PPN in vitro, in vivo, and across species, including humans.

Our findings established that every PPN neuron fired maximally at beta/gamma frequencies,[45] that all PPN neurons manifested beta/gamma frequency intrinsic membrane oscillations,[46] and that these oscillations were mediated by high threshold, voltage-dependent N- and/or P/Q-type calcium channels.[46] We found that these channels were distributed along the dendrites of PPN neurons,[47] and that some cells exhibited both N- and P/Q-type calcium channels, some had only N-type channels, and some had only P/Q-type channels.[48,49] We suggested that these three cell types represented neurons that were active in relation to waking and REM sleep (N + P/Q cells = "Wake-REM on" cells), only during REM sleep (N-only = "REM on" cells), or only during waking (P/Q-only = "Wake on" cells), respectively.[48–50] Fig. 3.1 shows the protocol used to identify these neurons. If the cell was a P/Q-only neuron (~25%), ramp-induced oscillations were not affected by the N-type channel blocker conotoxin, but were entirely blocked by agatoxin, a specific P/Q-type channel blocker. If the cell was an N-only neuron (~25%), ramp-induced oscillations were not affected by agatoxin but were completely blocked by conotoxin. If the cell was an N + P/Q

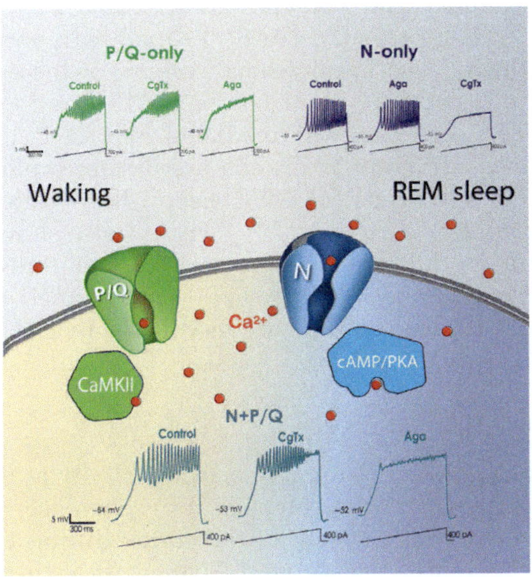

FIGURE 3.1 **Two types of calcium channels modulate waking versus REM sleep.** P/Q-type calcium (Ca^{2+}) channels (*green channel*) are modulated by CaMKII (*green sextagon*) intracellularly, while N-type channels (*blue channel*) are modulated by cAMP/PKA (*blue structure*) intracellularly. If the cell was a P/Q-only neuron (~25%), ramp-induced oscillations were not affected by the N-type channel blocker ω-Conotoxin-GVIA (CgTx), but were entirely blocked by ω-Agatoxin-IVA (Aga), a specific P/Q-type channel blocker (*green records*, top left). If the cell was a N-only neuron (~25%), ramp-induced oscillations were not affected by Aga but were completely blocked by CgTx (*purple records*, top right). If the cell was a N + P/Q neuron (~50%), CgTx only partially reduced oscillations and Aga blocked the remainder of the oscillations (*blue records*, bottom).

neuron (~50%), conotoxin only partially reduced oscillations and agatoxin blocked the remainder of the oscillations.

3.3.2 Two Intracellular Pathways

In freely moving animals, injections of glutamate into the PPN increased both waking and REM sleep,[51] but injections of NMDA, one glutamatergic receptor agonist, increased only waking,[52] while injections of kainic acid (KA), another glutamatergic receptor agonist, increased only REM sleep[53] (see also Chapter 2: Electrophysiological Correlates of the Sleep–Wake Cycle). These data indicated a differential control of waking versus REM sleep by different glutamatergic receptor subtypes. Protein kinase C, which modulates KA receptors, enhances N-type channel activity and has no effect on P/Q-type channel function[54]; however, CaMKII, which modulates NMDA receptors, was shown to modulate P/Q-type channel function.[55] These results suggest that the two calcium channel subtypes are modulated by different intracellular pathways, N-type by the cAMP/PK pathway, and P/Q-type via the CaMKII pathway. The implications from all of these results is that there is a "waking" pathway mediated by CaMKII and P/Q-type channels and a "REM sleep" pathway mediated by cAMP/PK and N-type channels; moreover, different PPN cells fire during waking (those with N + P/Q and only P/Q-type) or REM sleep (those with N + P/Q and only N-type) or both (N + P/Q).[21,26,46,50] Fig. 3.1 also shows the intracellular mechanisms modulating the two types of high threshold, voltage-dependent calcium channels, that is, P/Q-type channels are modulated by CaMKII, while N-type channels are modulated by cAMP/PKA.

3.3.3 Cortical Manifestations

Recent findings showed that gamma band activity at the level of the cortex during waking is characterized by coherence across distant regions, but gamma band activity in the cortex during REM sleep has an absence of coherence[56,57] (see also Chapter 4: Neurochemistry and Pharmacology of Sleep). Thus, since the brainstem is the origin of REM sleep drive, it is clear that the manifestation of gamma band activity during REM sleep at the level of the cortex begins in the brainstem. We assume that the manifestation of gamma band activity during waking is at least in part originating in the brainstem as well. Moreover, injections of the cholinergic agonist carbachol induced REM sleep with cataplexy that is characterized by decreased gamma band coherence in the cortex.[58] Therefore, this line of evidence suggests that not only do brainstem centers drive gamma band activity that is manifested in the cortical EEG, but during waking, brainstem-thalamic projections include coherence across regions, and during REM sleep, which is controlled by the subcoeruleus region (lesion of this region eliminates REM sleep, and injection of carbachol into it induces REM sleep), drives cortical EEG rhythms without coherence.[59,60]

Fig. 3.2 shows more of the intracellular and receptor control of the two calcium channels. Not shown on this figure are the effects of activation of, for example, a muscarinic cholinergic receptor typically acting through G protein coupling to phospholipase C, that in turn cleaves phospholipid phosphatidylinositol biphosphate into inositol triphosphate (IP3). IP3 is released and binds to IP3 receptors in the endoplasmic reticulum (ER) to release calcium (Ca^{2+}). One of the intracellular pathways activated involves CaMKII, which modulates P/Q-type calcium channels, and the other pathway involves cAMP/PKA, which modulates N-type calcium channels. The CaMKII/P/Q-type pathway mediates beta/gamma band activity during waking, while the cAMP/

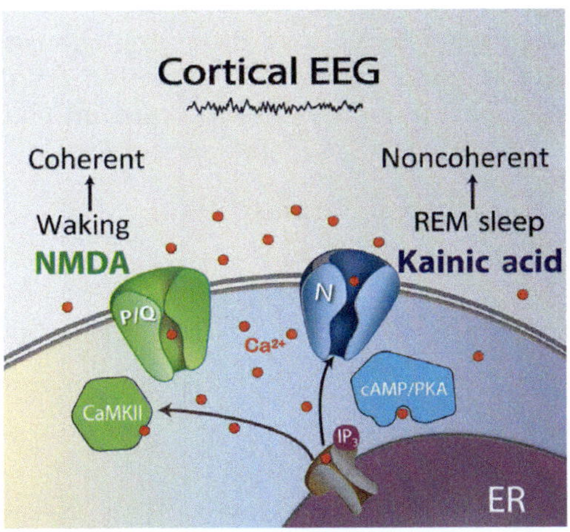

FIGURE 3.2 **Differential effects on cortical EEG of bottom-up gamma during waking versus REM sleep.** Upon receptor activation, IP3 is released and binds to IP3 receptors in the endoplasmic reticulum (ER) to release calcium (Ca^{2+}, *red dots*). One of the intracellular pathways activated involves CaMKII (*green sextagon*), which modulates P/Q-type calcium channels, and the other pathway involves cAMP/PKA (*blue structure*), which modulates N-type calcium channels. The CaMKII/P/Q-type pathway mediates beta/gamma band activity during waking, while the cAMP/PKA/N-type pathway mediates beta/gamma band activity during REM sleep. Moreover, activation of the NMDA receptor modulates P/Q-type channel activity, while activation of the kainic acid receptor modulates N-type channel activity. The gamma band oscillations generated by the PPN during waking ultimately lead to gamma activity in the cortical EEG with coherence across distant sites, while gamma activity during REM sleep leads to a similar fast cortical EEG but without coherence across distant sites (see text for relevant citations).

PKA/N-type pathway mediates beta/gamma band activity during REM sleep. Moreover, activation of NMDA receptors modulates P/Q-type channel activity, while activation of kainic acid receptors modulates N-type channel activity. Thus, gamma band oscillations generated by the PPN during waking ultimately lead to gamma activity in the cortical EEG with coherence across distant sites, while gamma activity during REM sleep leads to a similar fast cortical EEG but without coherence across distant sites.

3.4 BOTTOM-UP GAMMA

Fig. 3.3 depicts one origin of bottom-up gamma via sensory afferents that induce gamma oscillations in PPN neurons. These cells project to the parafascicular nucleus in the ILT, that also manifests N- and P/Q-type calcium channel-mediated gamma oscillations,[15] and these channels are located all along the dendrites of Pf neurons.[61] The ILT sends arousal-related input to the upper layers of the cortex, modulating low amplitude, high frequency EEG, and providing the context of sensory perception.[62] These suggestions allow a direction for determining, for example, the role of N-type and P/Q-type calcium channel expression

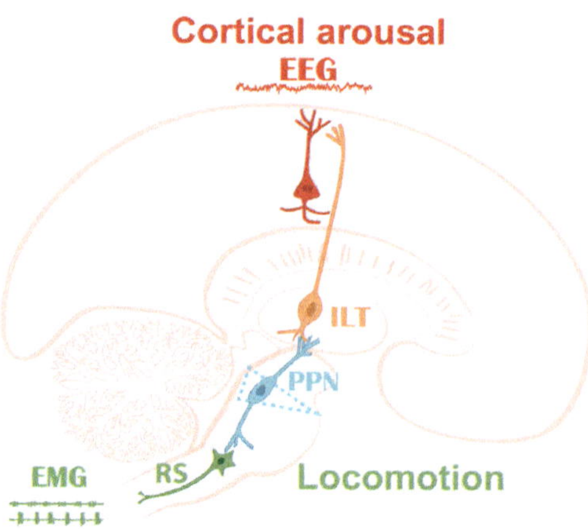

FIGURE 3.3 **Role of the PPN in cortical arousal and motor control.** The PPN sends ascending projections to the intralaminar thalamus (ILT) that in turn project to the upper layers of the cortex to help modulate low amplitude, high frequency EEG and cortical arousal. The PPN also sends descending projections to reticulospinal pathways to modulate the startle response (through inhibition of giant reticulospinal projections, not shown), as well as locomotion through excitation of medium reticulospinal (RS) neurons that in turn trigger spinal pattern generators to induce locomotion in fight-versus-flight responses.

in various neurological and psychiatric disorders, which has not been well explored. The role of intracellular pathways through CaMKII and waking, and cAMP/PKA and REM sleep, also require further attention since it is at intracellular sites that pathological states may better be treated. These and additional questions remain, but the discoveries described herein provide (1) a physiological mechanism that accounts for many of the arousal symptoms of a number of diseases, and (2) multiple novel directions for developing more effective treatments for these disorders.

Such efforts are important because disturbances in gamma band maintenance at the level of the PPN will disrupt a basic survival process. The ubiquitous generation of gamma band activity by all PPN neurons has been proposed, not to participate in the temporal process of sensory binding as in the cortex, but, as mentioned earlier, to underlie the more basic process of preconscious awareness.[26,63] That is, PPN gamma activity may participate in a process that provides the essential stream of information necessary for the formulation of many of our actions. Preconscious awareness is the state that allows us to reliably assess the world around us on a continuous basis, providing a unifying picture and promoting survival. Although this process does not involve selective attention to specific stimuli, it provides the background upon which we assess our environment and act upon it. Without this function, our very survival is at risk. Interruptions in this process, therefore, will markedly disrupt perception as well as planning of motor strategies, adding uncertainty to the wildly shifting world of the patient.

In addition, Fig. 3.3 shows the simultaneous activation of descending projections to reticulospinal pathways that influence the spinal cord in the form of postural changes in muscle tone resulting from the startle response, as well as trigger locomotor events in fight-versus-flight responses.[26] That is, the RAS is inextricably entwined with motor and postural control. Not only does the RAS modulate locomotion in fight-versus-flight responses, but also muscle tone in the atonia of REM sleep and the startle response. However, the involvement in motor control of the RAS, and of the PPN in particular, has been misunderstood.

3.5 LOCOMOTOR CONTROL

3.5.1 The Mesencephalic Locomotor Region

The mesencephalic locomotor region (MLR) was originally described in very specific terms. In the precollicular-postmamillary transected cat, stimulation using long duration (0.5–1.0 ms), low amplitude (10–100 μA) pulses delivered at specific frequencies (40–60 Hz) could elicit locomotion on a treadmill when the body was suspended on a sling.[64] Moreover, increasing stimulus amplitude would change stepping from a walk to a trot to a gallop, otherwise known as "controlled" locomotion on a treadmill. This is markedly different from the "unspecific" progression elicited by stimulation of multiple brain regions in an intact animal.[64,65] Such studies identify a locus independent of higher structures that could otherwise drive behavioral approach and escape. We investigated the neuronal substrates mediating this very specific effect and found that at least some low threshold sites in the rat and cat brains were located within the PPN when labeled for cholinergic neurons.[66] We further showed that the region could be chemically activated, eliminating the potential for electrical stimulation of fibers of passage.[67] However, we pointed out that there were other nearby sites that would lead to locomotion, especially when stimulated using different parameters.[65,68] We also pointed out that locomotion had to be "recruited" rather than induced, since it did not immediately follow the onset of stimulation, and that the region overlapped with elements of the RAS, especially the PPN.[68,69]

We repeatedly cast doubt on the specificity of this region, concluding that the MLR was not a "locomotion-specific" area, but rather a region that could be stimulated to induce rhythmic activity, that is, a "rhythmogenic" area.[65–69] In other words, stimulating at the natural frequency of this region would activate all of its efferent projections, including ascending ones projecting to the thalamus and basal ganglia modulating arousal, and descending ones to reticulospinal systems modulating posture and gait.[70,71] That is why PPN stimulation can simultaneously elicit cortical arousal and postural and locomotor changes.[72] It was not until much later that we discovered the reason for this natural frequency, the unique intrinsic oscillation beta/gamma frequency in these cells that could be optimally activated by stimulation at 40–60 Hz. That is, the fact that all PPN cells, regardless of transmitter type, manifest intrinsic membrane oscillations in the beta/gamma range mediated by high threshold calcium channels imparts on them a natural frequency of firing.[14,45,47]

Amazingly enough, a multitude of investigators have made assertions about the location of the MLR without labeling stimulation sites for cholinergic neurons,[73,74] or without carrying out actual "controlled" locomotion on a treadmill studies.[75,76] Moreover, using intact (nontransected) animals likely will test the many ways in which arousal can induce "unspecific" behavioral approach and escape in intact animals, but will not identify the site(s) essential for "controlled" locomotion.[73] Based on our suggestions over the years that stimulation of this region could be used therapeutically in PD,[65–72] there are now over 200 patients implanted with PPN deep brain stimulation (DBS) electrodes.

These studies have shown salutary effects on gait, posture, sleep, and even cognitive measures.[77] The most effective frequencies are not surprisingly in the 25–60 Hz (beta/gamma) range, but not at 10–20 Hz as recently purported.[73] Such low frequencies do not produce the range of beneficial effects of PPN DBS. The most effective frequencies (beta/gamma) are in keeping with the natural frequencies of firing of PPN neurons when they are activated.[14,45,46,48] Moreover, from a philosophical viewpoint, given the existence of spinal pattern generators for walking, why would there be a region high in the brainstem dedicated to locomotor control? We reiterate our position that the term "MLR" is outdated and should be retired.

3.5.2 Optogenetics

Optogenetic tools derived from microbial (type I) or animal (type II) opsins are widely used in a fast growing area of neuroscience that combines several animal models to target subpopulations of neurons within particular tissues/brain areas.[78] While this is a powerful technology, there are a number of caveats that need to be remembered. First, the combination of behavioral with optogenetics methods require in many cases the implantation of an optical fiber to deliver blue light capable of reaching subcortical structures. Recent studies show that blue light activates cells that do not express opsins.[79] Then there is the issue of latency. Although rapid resolution of opsins like channelrhodopsin or archaerhodopsin allows the study of fast interactions between cell groups, their association with behavioral events is not always rapid or direct. This makes it difficult to interpret multifactorial behavioral assays (i.e., sleep–wake rhythms, anxiety, fear, or depression tests) due to the long latency between stimulation and behavioral event. Stimulation to "recruit" locomotion typically has a latency of ~1 second, but optogenetic stimulation normally takes 10–20 seconds to induce an effect.[80] This link would be particularly remote in assessing whether animals sleep or awaken for longer periods or not.[81] Thirdly, the methodology employs the genetic introduction of the opsin using a vector, usually viral or transgenic, attached to a promoter that overexpresses the protein. Therefore, expression of opsins might permanently alter synaptic activity among brain nuclei by simply changing the lipid environment required for the normal membrane expression of channels and/or receptors.[82]

Many of these issues are particularly critical when studying the PPN. For example, in some groups of neurons like the PPN,[14,45,46,48] high frequency subthreshold oscillations require the use of a ramp-like, sustained depolarization that cannot be achieved using trains of light stimulation. Therefore, the kind of light stimulated activity induced may be far from that induced by the normal synaptic activation of calcium channels gradually recruiting intrinsic oscillations. Moreover, these channels are expressed all along the dendrites of PPN neurons.[47,61] The initial overexpression of the channelrhodopsin could interfere with the lipid rafts that organize, for example, high threshold, voltage-dependent channels.[83] This would markedly change the gamma band responsiveness of PPN neurons. Finally, since expressing these proteins in a cell could (1) alter protein expression, (2) opsins could induce heat that changes the cell, and (3) channel and pump activity can be altered,[84] the long-term insertion of these calcium leak channels could gradually change calcium intracellular concentrations, leading to abnormal responses, and it needs to be determined if such alterations can lead to cell death in the long run.

3.6 FUTURE DIRECTIONS

A field of study that has not received sufficient attention is that of neuroepigenetics. Does the manifestation of intrinsic gamma band oscillations lead to gene transcription? Is gene transcription different during waking versus REM sleep? Is gene transcription different when the P/Q-type channel/CaMKII pathway is activated compared with the N-type channel/cAMP/PKA pathway? What purpose would such transcription possess? Is it a mechanism for rebuilding the channels activated during a particular state? Or is it a mechanism that allows the consolidation of the organism's identity? That is, after repeated bouts of SWS during the night, do we need a reminder of who we are? During SWS, what are cortical cells doing? The EEG tells us that large ensembles of neurons are firing in unison during SWS, but during waking they are firing independently. What this means is that cortical columns, the constituent parts of the main mechanism regarded as responsible for perception and cortical function, are firing together during SWS. That is, there is no lateral inhibition since they are active synchronously. Thus, inputs to individual columns goes undetected; therefore, no contrast exists between inputs to different columns that would allow perception. That is why there is no perception during SWS. On the other hand, independent firing is restored within a second or so upon waking, and the business of perception is once again active. As mentioned above, increased blood flow in the brainstem and thalamus precedes that of the cortex on waking. Yet we immediately and seamlessly know who we are as soon as we awaken. Does gene transcription during waking help the daily reformulation of the self?

William James proposed that the "stream of consciousness" is "a river flowing forever through a man's conscious waking hours."[85] This pulsing stream persistently infiltrates our essence while waking and yet we often fail to pay it heed, letting much sensory information go unnoticed within the innermost confines of our mind. Once beckoned into awareness, we can actively pay tribute to a particular piece of sensory information: we become fully "conscious" of the information.[86] At the wellspring of this process is the RAS, a phylogenetically conserved area of the brain inundated by the continuous flow of internal (especially intrinsic oscillations) and external information (impinging on dendrites). By traversing the RAS, this information modulates sleep–wake cycles, the startle response, and fight-versus-flight responses, along with changes in muscle tone and locomotion. Accordingly, we speculate that activation of the RAS during waking induces coherent activity (through electrically coupled cells) and high-frequency oscillations (through P/Q-type calcium channel and subthreshold oscillations) to sustain gamma activity and support a persistent, reliable state for assessing our world.

Moreover, Penfield arrived at the conclusion that there was a mechanism for consciousness that was subcortical. He stated, "There is no place in the cerebral cortex where electrical stimulation will cause a patient to believe or to decide."[87] In addition, he emphasized that, while cortical seizures localized to specific regions elicit sensory or motor effects but maintain consciousness, *petit mal* seizures in "mesothalamic" (midbrain and thalamus) regions always eliminate consciousness. Based on these results, Penfield proposed the presence of a "centrencephalic integrating system" that fulfills the role of sensorimotor integration necessary for consciousness. Perhaps gene transcription in the RAS allows the maintenance of the core identity of the being from day to day.

Acknowledgments

Supported by NIH award P30 GM110702 from the IDeA program at NIGMS. EGR would also like to express profound gratitude to all of the Federal funding agencies, especially NIH and NSF, which have continuously funded his lab for the last 40 years. We also appreciate the assistance of S. Mahaffey with the figures in this chapter.

References

1. Garcia-Rill E, Kezunovic N, Hyde J, Beck P, Urbano FJ. Coherence and frequency in the reticular activating system (RAS). *Sleep Med Rev* 2013;**17**:227–38.
2. Thiagarajan TC, Lebedev MA, Nicolelis MA, Plenz D. Coherence potentials: loss-less, all-or-none network events in the cortex. *PLoS Biol* 2010;**8**:e1000278.
3. Garcia-Rill E, Heister DS, Ye M, Charlesworth A, et al. Electrical coupling: novel mechanism for sleep-wake control. *Sleep* 2007;**30**:1405–14.

4. Garcia-Rill E, Charlesworth A, Heister DS, Ye M, Hayar A. The developmental decrease in REM sleep: the role of transmitters and electrical coupling. *Sleep* 2008;**31**:673–90.
5. Heister DS, Hayar A, Charlesworth A, Yates C, Zhou J, et al. Evidence for electrical coupling in the SubCoeruleus (SubC) nucleus. *J Neurophysiol* 2007;**97**:3142–314.
6. Urbano FJ, Leznik E, Llinas R. Modafinil enhances thalamocortical activity by increasing neuronal electrotonic coupling. *Proc Natl Acad Sci USA* 2007;**104**:12554–9.
7. Evans WH, Boitano S. Connexin mimetic peptides: specific inhibitors of gap junctional intercellular communication. *Biochem Soc Trans* 2001;**29**:606–12.
8. He DS, Burt JM. Mechanism and selectivity of the effects of halothane on gap junction channel function. *Circ Res* 2000;**86**:1–10.
9. Boer DL, Henriksen SJ, Cravatt BF. Oleamide: an endogenous sleep-inducing lipid and prototypical member of a new class of biological signaling molecules. *Curr Pharm Des* 1998;**4**:303–14.
10. Murillo-Rodriguez E, Blanco-Centurion C, Sanchez C, Piomelli D, et al. Anandamide enhances extracellular levels of adenosine and induces sleep: an in vivo microdialysis study. *Sleep* 2003;**26**:943–7.
11. Gigout S, Louvel J, Kawasaki H, D'Antuono M, et al. Effects of gap junction blockers on human neocortical synchronization. *Neurobiol Dis* 2006;**22**:496–508.
12. Rozental R, Srinivas M, Spray DC. How to close a gap junction channel. *Methods Mol Biol* 2001;**154**:447–77.
13. Leznik E, Llinas R. Role of gap junctions in synchronized oscillations in the inferior olive. *J Neurophysiol* 2005;**94**:2447–56.
14. Kezunovic N, Urbano FJ, Simon C, Hyde J, Smith K, et al. Mechanism behind gamma band activity in the pedunculopontine nucleus (PPN). *Eur J Neurosci* 2011;**34**:404–15.
15. Kezunovic N, Hyde J, Simon C, Urbano FJ, Williams D, et al. Gamma band activity in the developing parafascicular nucleus (Pf). *J Neurophysiol* 2011;**107**:772–84.
16. Simon C, Kezunovic N, Williams DK, Urbano FJ, Garcia-Rill E. Cholinergic and glutamatergic agonists induce gamma frequency activity in dorsal subcoeruleus nucleus neurons. *Am J Physiol Cell Physiol* 2011;**301**:C327–35.
17. Llinas R, Yarom Y. Oscillatory properties of guinea-pig inferior olivary neurons and their pharmacological modulation: an in vitro study. *J Physiol* 1986;**376**:163–82.
18. Sperry RW. A modified concept of consciousness. *Psychol Rev* 1969;**76**:532–6.
19. Balkin TJ, Braun AR, Wesensten NJ, Jeffries K, Varga M, et al. The process of awakening: a PET study of regional brain activity patterns mediating the re-establishment of alertness and consciousness. *Brain* 2002;**125**:2308–19.
20. Steriade M. Cellular substrates of oscillations in corticothalamic systems during states of vigilance. In: Lydic R, Baghdoyan HA, editors. *Handbook of behavioral state control. Cellular and molecular mechanisms*. New York: CRC Press; 1999. p. 327–47.
21. Garcia-Rill E. Bottom-up gamma and stages of waking. *Med Hypothesis* 2017;**104**:58–62.
22. Moruzzi G, Magoun HW. Brain stem reticular formation and activation of the EEG. *Electroenceph Clin Neurophysiol* 1949;**1**:455–73.
23. Watson RT, Heilman KM, Miller BD. Neglect after mesencephalic reticular formation lesions. *Neurology* 1974;**24**:294–8.
24. Vanderwolf CH. What is the significance of gamma wave activity in the pyriform cortex? *Brain Res* 2000;**877**:125–33.
25. Vanderwolf CH. Are neocortical gamma waves related to consciousness? *Brain Res* 2000;**855**:217–24.
26. Garcia-Rill E. *Waking and the reticular activating system*. New York: Academic Press; 2015. p. 330.
27. Eckhorn R, Bauer R, Jordan W, Brosch M, Kruse W, et al. Coherent oscillations: a mechanism of feature linking in the visual system? *Biol Cybern* 1988;**60**:121–30.

28. Gray CM, Singer W. Stimulus-specific neuronal oscillations in orientation columns of cat visual cortex. *Proc Natl Acad Sci USA* 1989;**86**:1698–702.
29. Philips S, Takeda Y. Greater frontal-parietal synchrony at low gamma-band frequencies for inefficient then efficient visual search in human EEG. *Int J Psychophysiol* 2009;**73**:350–4.
30. Palva S, Monto S, Palva JM. Graph properties of synchronized cortical networks during visual working memory maintenance. *Neuroimage* 2009;**49**:3257–68.
31. Llinas R, Leznik E, Urbano FJ. Temporal binding via cortical coincidence detection of specific and nonspecific thalamocortical inputs: a voltage-dependent dye-imaging study in mouse brain slices. *Proc Natl Acad Sci USA* 2002;**99**:449–54.
32. Steriade M, Llinás R. The functional states of the thalamus and the associated neuronal interplay. *Physiol Rev* 1988;**68**:649–742.
33. Colgin LL, Moser EI. Gamma oscillations in the hippocampus. *Physiology (Bethesda)* 2010;**25**:319–29.
34. Jenkinson N, Kuhn AA, Brown P. Gamma oscillations in the human basal ganglia. *Exp Neurol* 2013;**245**:72–6.
35. Middleton SJ, Racca C, Cunningham MO, Traub RD, Monyer H, et al. High-frequency network oscillations in cerebellar cortex. *Neuron* 2008;**58**:763–74.
36. Sakai K, El Mansari M, Jouvet M. Inhibition by carbachol microinjections of presumptive cholinergic PGO-on neurons in freely moving cats. *Brain Res* 1990;**527**:213–23.
37. Steriade M, Paré D, Datta S, Oakson G, Curro Dossi R. Different cellular types in mesopontine cholinergic nuclei related to ponto-geniculo-occipital waves. *J Neurosci* 1990;**10**:2560–79.
38. Kayama Y, Ohta M, Jodo E. Firing of 'possibly' cholinergic neurons in the rat laterodorsal tegmental nucleus during sleep and wakefulness. *Brain Res* 1992;**569**:210–20.
39. Datta S, Siwek DF. Single cell activity patterns of pedunculopontine tegmentum neurons across the sleep-wake cycle in the freely moving rats. *J Neurosci Res* 2002;**70**:79–82.
40. Datta S, Siwek DF, Stack EC. Identification of cholinergic and non-cholinergic neurons in the pons expressing phosphorylated cyclic adenosine monophosphate response element-binding protein as a function of rapid eye movement sleep. *Neuroscience* 2009;**163**:397–414.
41. Boucetta S, Cisse Y, Mainville L, Morales M, Jones BE. Discharge profiles across the sleep-waking cycle of identified cholinergic, GABAergic, and glutamatergic neurons in the pontomesencephalic tegmentum of the rat. *J Neurosci* 2014;**34**:4708–27.
42. Steriade M, Datta S, Pare D, Oakson G, Curro Dossi RC. Neuronal activities in brainstem cholinergic nuclei related to tonic activation processes in thalamocortical systems. *J Neurosci* 1990;**10**:2541–59.
43. Fraix V, Bastin J, David O, Goetz L, Ferraye M, et al. Pedunculopontine nucleus area oscillations during stance, stepping and freezing in Parkinson's disease. *PLoS One* 2008;**8**:e83919.
44. Goetz L, Piallat B, Bhattacharjee M, Mathieu H, David O, et al. The primate pedunculopontine nucleus region: towards a dual role in locomotion and waking state. *J Neural Transm* 2016;**123**:667–78.
45. Simon C, Kezunovic N, Ye M, Hyde J, Hayar A, et al. Gamma band unit activity and population responses in the pedunculopontine nucleus. *J Neurophysiol* 2010;**104**:463–74.
46. Urbano FJ, Kezunovic N, Hyde J, Simon C, Beck P, et al. Gamma band activity in the reticular activating system (RAS). *Front Neurol* 2012;**3**(6):1–16.
47. Hyde J, Kezunovic N, Urbano FJ, Garcia-Rill E. Spatiotemporal properties of high speed calcium oscillations in the pedunculopontine nucleus. *J Appl Physiol* 2013;**115**:1402–14.

48. Luster B, D'Onofrio S, Urbano FJ, Garcia-Rill E. High-Threshold Ca2+ channels behind gamma band activity in the pedunculopontine nucleus (PPN). *Physiol Rep* 2015;**3**:e12431.
49. Luster B, Urbano FJ, Garcia-Rill E. Intracellular mechanisms modulating gamma band activity in the pedunculopontine nucleus (PPN). *Physiol Rep* 2016;**4**:e12787.
50. Garcia-Rill E, Kezunovic N, D'Onofrio S, Luster B, Hyde J, et al. Gamma band activity in the RAS-intracellular mechanisms. *Exp Brain Res* 2014;**232**:1509–22.
51. Datta S, Spoley EE, Patterson EH. Microinjection of glutamate into the pedunculopontine tegmentum induces REM sleep and wakefulness in the rat. *Am J Physiol Reg Integ Comp Physiol* 2001;**280**:R752–9.
52. Datta S, Patterson EH, Spoley EE. Excitation of the pedunculopontine tegmental NMDA receptors induces wakefulness and cortical activation in the rat. *J Neurosci Res* 2002;**66**:109–16.
53. Datta S. Evidence that REM sleep is controlled by the activation of brain stem pedunculopontine tegmental kainate receptor. *J Neurophysiol* 2002;**87**:1790–8.
54. Stea A, Soomg TW, Snutch TP. Determinants of PKC-dependent modulation of a family of neuronal Ca^{2+} channels. *Neuron* 1995;**15**:929–40.
55. Jiang X, Lautermilch NJ, Watari H, Westenbroek RE, Scheuer T, et al. Modulation of $Ca_v2.1$ channels by Ca^+/calmodulin-dependent kinase II bound to the C-terminal domain. *Proc Natl Acad Sci USA* 2008;**105**:341–6.
56. Castro S, Falconi A, Chase M, Torterolo P. Coherent neocortical 40-Hz oscillations are not present during REM sleep. *Eur J Neurosci* 2013;**37**:1330–9.
57. Cavelli M, Castro S, Schwartzkopf N, Chase M, Falconi A, et al. Coherent cortical oscillations decrease during REM sleep in the rat. *Behav Brain Res* 2015;**281**:318–25.
58. Torterolo P, Castro-Zaballa S, Cavelli M, Chase M, Falconi A. Neocortical 40 Hz oscillations during carbachol-induced rapid eye movement sleep and cataplexy. *Eur J Neurosci* 2015;**281**:318–25.
59. Garcia-Rill E, D'Onofrio S, Mahaffey S. Bottom-up gamma: the pedunculopontine nucleus and reticular activating system. *Transl Brain Rhythm* 2016;**1**:49–53.
60. Garcia-Rill E, Virmani T, Hyde J, D'Onofrio S, Mahaffey S. Arousal and the control of perception and movement. *Curr Trends Neurol* 2016;**10**:53–64.
61. Hyde J, Kezunovic N, Urbano FJ, Garcia-Rill E. Visualization of fast calcium oscillations in the parafascicular nucleus. *Pflugers Arch* 2013;**465**:1327–40.
62. Llinas R, Leznik E, Urbano FJ. Temporal binding via cortical coincidence detection of specific and nonspecific thalamocortical inputs: a voltage-dependent dye-imaging study in mouse brain slices. *Proc Natl Acad Sci USA* 2002;**99**:449–54.
63. Urbano FJ, D'Onofrio S, Luster B, Hyde J, Bisagno V, et al. Pedunculopontine nucleus gamma band activity-preconscious awareness, waking, and REM sleep. *Front Sleep Chronobiol* 2014;**5**:210.
64. Shik ML, Severin FV, Orlovskii GN. Control of walking and running by means of electric stimulation of the midbrain. *Biofizika* 1966;**11**:659–66.
65. Garcia-Rill E. The basal ganglia and the locomotor regions. *Brain Res Rev* 1986;**11**:47–63.
66. Garcia-Rill E, Houser CR, Skinner RD, Smith W, Woodward DJ. Locomotion-inducing sites in the vicinity of the pedunculopontine nucleus. *Brain Res Bull* 1987;**18**:731–8.
67. Garcia-Rill E, Skinner RD, Fitzgerald JA. Chemical activation of the mesencephalic locomotor region. *Brain Res* 1985;**330**:43–54.
68. Garcia-Rill E, Skinner RD. Modulation of rhythmic function in the posterior midbrain. *Neuroscience* 1988;**17**:639–54.
69. Skinner RD, Garcia-Rill E. Brainstem modulation of rhythmic functions and behaviors. In: Klemm WR, Vertes RP, editors. *Brainstem mechanisms of behavior*. New York: John Wiley & Sons; 1990. p. 419–45.

70. Garcia-Rill E. The pedunculopontine nucleus. *Prog Neurobiol* 1991;**36**:363–89.
71. Reese NB, Garcia-Rill E, Skinner RD. The pedunculopontine nucleus-auditory input, arousal and pathophysiology. *Prog Neurobiol* 1995;**47**:105–33.
72. Garcia-Rill E, Reese NB, Skinner RD. Arousal and locomotion: from schizophrenia to narcolepsy. In: Holstege G, Saper C, editors. *The emotional motor system, Prog Brain Res*, vol. 107. 1996. p. 417–34.
73. Noga B, Sanchez F, Villamil L, O'Toole C, Kasicki S, et al. LFP oscillations in the mesencephalic locomotor region during voluntary locomotion. *Front Neural Circuits* 2017;**11**:34.
74. Takakusaki K, Habaguchi T, Ohtinata-Sugimoto J, Saitoh K, Sakamoto T. Basal ganglia efferents to the brainstem centers controlling postural muscle tone and locomotion: a new concept for understanding motor disorders in basal ganglia dysfunction. *Neuroscience* 2003;**119**:293–308.
75. Mena-Segovia J, Bolam P. Rethinking the pedunculopontine nucleus: from cellular organization to function. *Neuron* 2017;**94**:7–18.
76. Sherman D, Fuller P, Marcus J, Yu J, Zhang P, et al. Anatomical location of the mesencephalic locomotor region and its possible role in locomotion, posture, cataplexy, and Parkinsonism. *Front Neurol* 2015;**6**:140.
77. Garcia-Rill E, Hyde J, Kezunovic N, Urbano FJ, Petersen E. The physiology of the pedunculopontine nucleus-implications for deep brain stimulation. *J Neural Transm* 2014;**122**:225–35.
78. Fenno L, Yizhar O, Deisseroth K. The development and application of optogenetics. *Ann Rev Neurosci* 2011;**34**:389–412.
79. Cheng KP, Kiernan EA, Eliceiri KW, Williams JC, Watters JJ. Blue light modulates murine microglial gene expression in the absence of optogenetic protein expression. *Sci Rep* 2016;**6**:21172.
80. Garcia-Rill E, Mahaffey S, Hyde J, Urbano FJ. Bottom-up gamma maintenance in various disorders. In: Neurobiology of disease: pedunculopontine nucleus deepbrain stimulation, *Neurobiol Dis* 2018. Jan 17. pii: S0969-9961(18)30016-0. http://dx.doi.org/10.1016/j.nbd.2018.01.010. [Epub ahead of print].
81. Scammell T, Arrigoni E, Lipton J. Neural circuitry of wakefulness and sleep. *Neuron* 2018;**93**:747–65.
82. Brady J, Rich T, Le X, Stafford K, Fowler C, et al. Functional role of lipid raft microdomains in cyclic nucleotide-gated channel activation. *Mol Pharmacol* 2004;**65**:503–11.
83. Guéguinou M, Gambade A, Félix R, Chantôme A, Fourbon Y, et al. Lipid rafts, KCa/ClCa/Ca^{2+} channel complexes and EGFR signaling: novel targets to reduce tumor development by lipids? *Biochim Biophys Acta* 2015;**1848**(10 Pt B):2603–20.
84. Allen B, Singer AC, Boyden ES. Principles of designing interpretable optogenetic behavior experiments. *Learn Mem* 2015;**22**:232–8.
85. James W. *The principles of psychology*, vol. 1. New York: H. Holt; 1890. p. 225.
86. Civin M, Lombardi KL. The preconscious and potential space. *Psychoanal Rev* 1990;**77**:573–85.
87. Penfield W. *The mystery of the mind*. Princeton, NJ: Princeton University Press; 1970. p. 157.

CHAPTER

4

Neurochemistry and Pharmacology of Sleep

Pablo Torterolo[1], *Jaime M. Monti*[2] *and Giancarlo Vanini*[3]

[1]Laboratory of Sleep Neurobiology, Department of Physiology, School of Medicine, Universidad de la República, Montevideo, Uruguay [2]Department of Pharmacology and Therapeutics, School of Medicine, Clinics Hospital, Universidad de la República, Montevideo, Uruguay [3]Department of Anesthesiology, University of Michigan, Ann Arbor, MI, United States

OUTLINE

DOI: https://doi.org/10.1016/B978-0-12-816430-3.00004-X

ABBREVIATIONS

2-AG 2-archidonylglycerol
AEA anandamide
BF basal forebrain
CB1 cannabinoid receptor 1
CNS central nervous system

DAT	dopamine active transporter
DORA	dual orexin receptor antagonist
DR	dorsal raphe nucleus
DREADD	designer receptors exclusively activated by designer drugs
EDS	excessive daytime sleepiness
EEG	electroencephalogram
IL-1	interleukin-1
LC	locus coeruleus
LDT	latero-dorsal tegmental nuclei
MCH	melanin-concentrating hormone
NO	nitric oxide
NPO	nucleus pontis oralis
NREM	nonrapid eye movement
POA	preoptic area
PPT	peduncle-pontine tegmental nuclei
RBD	REM sleep behavioral disorder
REM	rapid eye movement
SCN	suprachiasmatic nucleus
TNF	tumor necrosis factor
vlPAG	ventrolateral periaqueductal gray
VLPO	ventrolateral region of the preoptic area
VTA	ventral tegmental area
W	wakefulness

4.1 INTRODUCTION

Knowledge of the neurochemical basis of sleep provides the clinician with the foundations to understand sleep pathology and pharmacological strategies for its treatment. Sleep pathologies are characterized by the lack of quantity or necessary quality of sleep (insomnia), or by sleep excess (hypersomnia). There are also syndromes characterized by the presence of abnormal behaviors during sleep (parasomnias) as well as circadian maladjustments. In addition, sleep related breathing and movement disorders are other types of sleep pathologies.[1] Finally, the modern way of living also challenges the physiological sleep needs, and has engendered an epidemic of chronic partial sleep deprivation.

In many of these conditions, a neurochemical imbalance has been demonstrated, and the pharmacological treatment strategy is to recover the neurochemical equilibrium. In this work, we review the state of the art of the neurochemical basis of sleep and wakefulness. With this foundation in mind, we also discuss the neurobiological basis of the current pharmacological approaches to treat paradigmatic sleep disturbances.

4.2 SLEEP AND WAKEFULNESS

In humans (and mammals in general), three behavioral states can be distinguished: wakefulness (W), nonrapid eye movement (NREM) sleep (also called slow wave sleep), and rapid eye movement (REM) sleep.

These behavioral states can be recognized by means of polysomnography, which consists of the simultaneous recording of various physiological parameters such as electroencephalogram (EEG), electromyogram, and electrooculogram.

Consciousness (awareness) is the cognitive counterpart of W. The EEG recording of W is marked by the presence of high frequency and low voltage oscillations (cortical activation). In the falling asleep process, adults enter into NREM sleep. In humans, three NREM sleep phases are recognized: N1, N2, and N3, according to the depth of the state. N1 is the transitional stage from W, while N2 is characterized by the presence of sleep spindles and K-complexes. The presence of low frequency (0.5–4 Hz, delta oscillations) of high amplitude waves characterizes the EEG during N3.[2] Although dreams can occur during light NREM sleep, oneiric activity is scarce during N3.[3–5]

REM sleep (also called stage R), is a deep sleep stage even though the EEG is similar to that of W; hence, it is also called "paradoxical" sleep. Dreams occur mainly during this sleep state. REM sleep is characterized by rapid eye movements, muscle atonia, and phasic changes in autonomic activity. REM sleep occupies 20%–25% of total sleep in adults and occurs approximately 90 minutes after the onset of sleep, a parameter known as REM sleep latency.[2] This latency decreases in pathological conditions such as narcolepsy and major depression.[6,7] REM sleep, as well as active W, is also characterized by theta (4–8 Hz) electrographic rhythm in the hippocampus; this signal is easily observed in rodents.

Nighttime sleep in humans is characterized by the presence of four to five sleep cycles. They comprise the period between the onset of sleep until the end of the first episode of REM sleep, or the period from the end of an episode of REM sleep to the end of the subsequent REM sleep episode. The average duration of sleep cycles is approximately 90 minutes in adults.[2]

The polysomnographic sleep patterns described above differ in the newborn; sleep matures along with the development of the brain.[8,9]

4.3 MOST FUNCTIONS ARE MODIFIED DURING SLEEP

When passing from one behavioral state to the other, the following main functions are deeply affected:

1. Cognitive functions. As mentioned earlier, we are conscious during W but during deep NREM sleep (N3) cognitive activity (dreams) is

scarce. On the contrary, dreams are the cognitive counterpart of REM sleep[3–5]; this "bizarre" cognitive activity has been equalized with a natural psychosis. In fact, according to Hobson (1997), "dreaming is, by definition, a psychosis."[10]

Cognitive activities (consciousness and dreams), and the different EEG rhythms that support these functions, are mainly generated by the activity of cortical and thalamic neuronal networks, which are mutually interconnected. Thalamic neurons have a complex electrophysiology that allows them to operate differently according to their level of polarization.[11] When hyperpolarized, the thalamic neurons that project to the cortex (thalamocortical neurons) oscillate at low frequency (0.5–4 Hz), and tend to block the sensory information that travels toward the cortex. This "oscillatory mode" of function synchronizes the cortical neurons and, accompanied by other phenomena of cortical origin, generates the slow waves of NREM sleep.[12,13] On the contrary, when these neurons are relatively depolarized, they enter in the "tonic mode" of function. In this condition, the thalamocortical neurons transmit sensory information toward the cortex in a reliable way. This mode of function occurs during W and REM sleep. Therefore, the thalamus is critical for the generation of part of the slow waves of NREM sleep. Moreover, the reticular nucleus of the thalamus is the site of generation of the sleep spindles that characterize N2 sleep.[13,14] When the thalamus is lesioned, as occurs in the "fatal familial insomnia," the generation of these electrographic signs is suppressed and deep NREM sleep does not occur.[15]

2. Motor functions. Whereas movements and high muscle tone characterize W, the tone of the somatic muscles decreases during NREM sleep. REM sleep is distinguished because of the deep muscle atonia, mainly of antigravity muscles. Due to the brain being very active during REM sleep, the atonia caused by the hyperpolarization of motoneurons is a protective mechanism that prevents acting out the dreams. The main respiratory muscle, the diaphragm, is not affected by the atonia. However, respiratory activity (frequency and tidal volume) is deeply affected during the sleep–waking cycle.[16]
3. Visceral functions. Autonomic and endocrine activity is highly dependent on the behavioral state; hence there are profound modifications on visceral activity during sleep.[17] For example, a hallmark of NREM sleep is the increase in tonic parasympathetic activity; this fact determines heart-rate adjustments during this state.[18]

4.4 SYNOPSIS OF WAKING AND SLEEP MECHANISMS

Thalamocortical, premotor/motor, autonomic, and hypothalamic endocrine neuronal networks are highly affected during the waking–sleep cycle, determining the functional aforesaid adjustments. However, the "primary engines" that determine changes in these neuronal networks are distributed in what are known as W-promoting (activating) and sleep-promoting systems. The location of the main components of these systems is shown in Fig. 4.1. The waking and the NREM sleep promoting neuronal networks have reciprocal inhibition (Fig. 4.2). The pattern of activity in the NREM sleep and W promoting regions has been modeled as a flip-flop electrical switch[19]; that is, periodic activation of either the W-promoting or NREM-promoting neuronal groups, with reciprocal inhibition between them. Interestingly, the structures that are critical for REM sleep generation mostly overlap with the neuronal networks that promote W.

The sleep–wake cycle is regulated by a circadian component (process C) and an homeostatic component (process S).[20] In humans, the propensity to sleep increases during the night (determined by process C) and increases proportionally to the duration of the episode of W (determined by the process S). Hence, the tendency to sleep is reproduced by mathematical models that include these two processes.

The circadian master-clock, the suprachiasmatic nucleus (SCN) of the hypothalamus, is the key component of the process C.[21] Both gene expression and electrical activity of the neurons of this nucleus (mostly GABAergic) oscillate with a 24-hour period, in the absence of environmental cues.[22] The SCN receives information about the environmental light directly from the retina. In diurnal animals, such as humans, SCN neurons promote W, probably by facilitating and inhibiting the activity of the W-promoting and NREM-promoting neurons, respectively (Fig. 4.2). Experimental evidence suggests that the pathway from the SCN for W and sleep regulation is mediated by the subparventricular zone and the dorsomedial hypothalamic nuclei.[23,24]

It is considered that the homeostatic process S is initiated by substances, such as adenosine, that are released into specific areas of the central nervous system (CNS) during W and promote NREM sleep[25,26] (Fig. 4.2).

W is enhanced during fasting, whereas sleep is favored after food intake.[27,28] In fact, metabolic signals affect sleep. For example, glucose activates the NREM-promoting neurons of the ventrolateral preoptic area (VLPO) and decreases the activity of the W-promoting hypocretinergic neurons.[29,30]

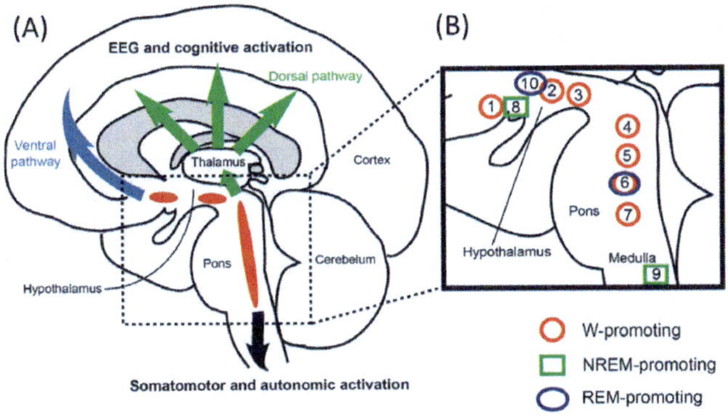

FIGURE 4.1 Scheme of the wakefulness and sleep-promoting systems. (A) Activating systems. The approximate anatomical location of the activating systems is shown in *red*. Activating systems are a heterogeneous group of neurons that use different neurotransmitters. The dorsal ascending activating pathways (that reach the thalamus influencing this on the cerebral cortex) and ventral (directly reaching the cerebral cortex) are shown in *green* and *blue*, respectively. Through these routes, the activating systems modulate the level of vigilance, cognitive functions, and EEG activity. In turn, descending projections regulate motor, respiratory, and autonomic activity. By projections to the hypothalamus (not shown) the activating systems also regulate the endocrine activity. (B) Outline of the main wakefulness, REM, and NREM sleep promoting nuclei. The neurons that form the activating systems (in *red*) are found in the basal forebrain, posterolateral hypothalamus, and mesopontine reticular formation, and use different neurotransmitters. The numbers identify different neuronal groups (approximate location): (1) Basal forebrain, cholinergic neurons. (2) Posterolateral hypothalamus, hypocretinergic neurons. (3) Tuberomammillary nucleus of the hypothalamus, histaminergic neurons. (4) Ventral tegmental area and substantia nigra, dopaminergic neurons. (5) Dorsal and medial raphe nucleus, serotonergic neurons. (6) Latero-dorsal and peduncle-pontine tegmental nucleus (LDT-PPT) neurons, cholinergic neurons. (7) Locus coeruleus, noradrenergic neurons. In *green*, the NREM sleep promoting regions are indicated; the preoptic region of the hypothalamus (8) and the parafacial zone (9). The cholinergic neurons of LDT-PPT (6) and the MCHergic neurons of the posterolateral hypothalamus (10) that promote REM sleep are shown in *blue*. The MCHergic neurons (10) also promote NREM sleep.

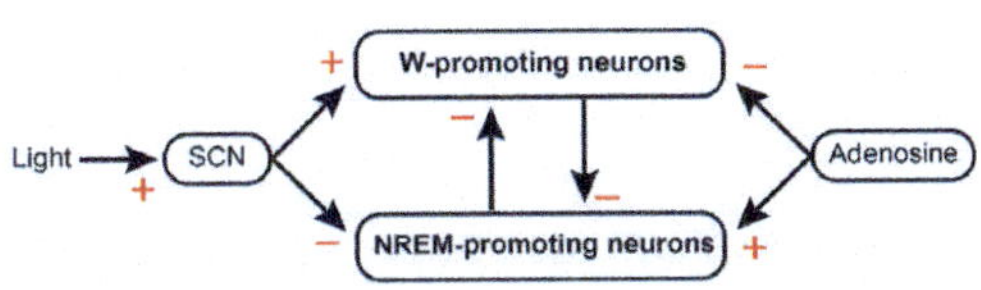

FIGURE 4.2 Scheme of the reciprocal inhibition between the wakefulness-promoting and NREM-promoting neuronal networks, and the regulation by the suprachiasmatic nucleus (SCN, process C) and by adenosine (process S). W, wakefulness. +, −, activation and inhibition, respectively.

Since Moruzzi and Magoun's classical report it is considered that sensory stimulation promotes W by activation of the W-promoting system[31]; however, repetitive sensory stimulation may also promote sleep.[32–34]

It is important to note that although both the standard EEG and behavior may denote W, local populations of neurons in the cortex may show a NREM sleep firing profile with negative consequences for performance. This phenomenon is called local seep, and local sleep homeostasis may be involved in this process.[35,36] In fact, a recent study has shown that local cortical activity of sleeping mice could be converted to the EEG profile of W by imposing a change in the extracellular ion composition.[37]

Very little is known about the neuronal bases involved in the transition from NREM sleep to REM sleep. It has been proposed that the mesopontine peribrachial region, which is activated during NREM sleep, could be involved in promoting the transition between these two sleep states.[38]

4.5 NEURONAL MODELS OF SLEEP AND WAKEFULNESS

Early studies, mainly performed in cats, pointed to the cholinergic and monoaminergic neuronal networks as the main structures responsible for the generation and maintenance of W (and REM sleep, see below). More recent studies in rodents, which employed powerful optogenetic and chemogenetic tools, have suggested that glutamatergic and GABAergic neurons, whose soma are widespread in different regions of the brain, are the main elements for the generation of behavioral states. The authors who support this model consider monoaminergic, cholinergic, and also neuropeptidergic neurons as regulators, but not as the main engine for the generation of the behavioral states. In this work, we will review the main structures and neurotransmitters involved in W and sleep.

4.6 WAKEFULNESS-PROMOTING NEURONAL NETWORKS AND NEUROTRANSMITTERS

The critical neural systems for the generation and maintenance of W are known as W-promoting or activating systems.[39] These are composed of neuronal groups located mainly within the mesopontine reticular formation, posterolateral hypothalamus, and basal forebrain (BF). All of them project to the thalamus and/or cerebral cortex, producing cortical activation. The ascending projections of these neuronal groups are organized in a dorsal and ventral pathway. The dorsal pathway goes through the thalamus and is mainly involved in the generation of the content (cognitive counterpart) of W, while the ventral pathway goes through the lateral hypothalamus and BF toward the cortex and has been related to the generation of W per se.[40] The activating systems are summarized in Fig. 4.1.

There is an important anatomical and functional interrelation between the neuronal groups that are part of the activating systems. Because there are different types of W, different activating neuronal groups participate more in the generation of one type of W than another. For example, noradrenergic neurons (see below) would have a preponderant role during W that supports attentional processes, serotonergic neurons would sustain W that is accompanied by stereotyped and automatic motor activity, while dopaminergic and hypocretinergic neurons would maintain a waking state with high motivation.

The firing rate of the W-promoting neurons and the release of their neurotransmitters into the synaptic cleft tend to be maximal during W and decreases during NREM sleep (Table 4.1). Since REM sleep is characterized by activation of the EEG associated with oneiric activity, it is expected that part of the activating systems remain active in this state. In fact, cholinergic neurons increase their activity during this behavioral state (Table 4.1).

The activity of these neuronal networks that employ different neurotransmitters promotes a synaptic neurochemical milieu that determines the promotion of W. Below, we will review the main neurotransmitters and neurochemical systems that are involved in the generation of W.

4.6.1 Noradrenaline

In the CNS, noradrenaline is found almost exclusively in the neurons of the locus coeruleus (LC), located in the dorsolateral mesopontine region (Fig. 4.1). These noradrenergic neurons have a "W-ON," "REM-OFF" profile[41] (Table 4.1). This profile is in accordance with the pattern of synaptic release of noradrenaline into the cerebral cortex.[42] During W these neurons increase their firing rate in response to a novel stimulus, but this response is reduced after habituation,[43] which led to the proposal that this neuronal group regulates attentional processes. Experiments in mice have shown that optogenetic stimulation of noradrenergic neurons rapidly awakes the animals, while inhibition of these neurons promotes transition into NREM sleep and reduces W.[44]

TABLE 4.1 Main Neurotransmitters/Neuromodulators That Have an Active Role in the Generation of Wakefulness and Sleep

Neurotransmitters/ neuromodulators	Localization of the neurons	Activity profile
	WAKEFULNESS-PROMOTING	
Hipocretins/Orexins	Posterolateral hypothalamus	W-on
Acetylcholine	LDT-PPT and BF	W/REM-on
Serotonin	Dorsal raphe nucleus	W-on
Noradrenaline	Locus coeruleus	W-on
Dopamine	VTA and SN	Burst discharge during W and REM[a]
Histamine	TMN	W-on
GABA[b]	Nucleus pontis oralis	W-on
Glutamate[b]	Nucleus pontis oralis	W-on
	NREM-SLEEP PROMOTING	
Melanin-concentrating hormone (MCH)	Posterolateral hypothalamus	NREM-on
Adenosine	c	c
GABA[b]	POA	NREM-on
	REM-SLEEP PROMOTING	
Acetylcholine	LDT-PPT and BF	W/REM-on and REM-on
Melanin-concentrating hormone (MCH)	Posterolateral hypothalamus	REM-on
Glutamate[b]	Nucleus pontis oralis	W/REM-on

[a] *During the sleep–wake cycle, the pattern but not the frequency of discharge changes.*
[b] *The role of GABAergic and glutamatergic neurons depends on their location; only the main examples were included in the list.*
[c] *The extracellular release of adenosine increases with prolonged wakefulness, but the origin of adenosine (metabolic, glial, or released by neurons as neurotransmitter) is still not clear.*
BF, basal forebrain; *LDT-PPT*, latero-dorsal and peduncle-pontine tegmental nucleus; *POA*, preoptic area of the hypothalamus; *SN*, substantia nigra; *TMN*, tuberomammillar nucleus of the hypothalamus; *VTA*, ventral tegmental area; *W*, wakefulness.

Interestingly, when rats with LC lesions were introduced into a socially and physically complex environment, they had less W than controls.[45]

Pharmacological note. Alpha-1 adrenergic antagonists such as prazosin facilitate the generation of sleep, and alpha-2 agonists such as dexmedetomidine, which inhibit the activity of LC neurons, are used as sedatives.[46,47]

4.6.2 Dopamine

Dopamine is found in neurons whose soma is located in mesencephalic substantia nigra pars compacta and ventral tegmental area (VTA) (Fig. 4.1). VTA dopaminergic neurons project to the prefrontal cortex, while both neuronal groups project to the striatum.[48] Dopaminergic neurons do not change their firing rate during the sleep–wake cycle, but their temporal pattern of discharge is highly modified (Table 4.1). During W, dopaminergic neurons discharge high frequency "bursts or trains" in response to a motivational stimulus, which produces a large release of dopamine into the synaptic space.[49,50] During REM sleep, this firing pattern is also observed in dopaminergic neurons of the VTA; the dopaminergic release in the nucleus accumbens (ventral striatum) also increases during REM sleep.[51,52] Interestingly, dopamine active transporter (DAT) knockout mice showed a significant increase of W and a reduction of NREM sleep, probably related with an increase in dopamine in the synaptic cleft.[53] Optogenetic activation of VTA dopaminergic neurons has been found to induce and maintain W during a period of high sleep pressure, while chemogenetic inhibition of these neurons during the dark phase, when W predominates, induces NREM sleep.[54]

Lu and coworkers have described a group of dopaminergic neurons within the ventral periaqueductal gray that may also be important in the generation and maintenance of W.[55] This group of neurons is probably functionally similar to the dopaminergic neurons of the rostral region of the dorsal raphe (DR); in fact, a recent study has shown that optogenetic activation of these neurons promotes W.[56]

Pharmacological note. Dopamine agonists and antagonists have been shown to modify sleep variables in preclinical and clinical studies.[57,58] In this respect, several second-generation antipsychotic drugs including clozapine, olanzapine, and paliperidone that block the dopamine D_2 receptor cause a significant reduction of sleep latency, and an increase of total sleep time and N2 when administered to schizophrenia patients.[59,60]

The piperidine derivative methylphenidate has been approved for the treatment of attention deficit hyperactivity disorder in children and adults, and as a second-line treatment for narcolepsy in adults.[61] This drug is also useful for the treatment of idiopathic hypersomnia and excessive daytime sleepiness (EDS) related to Parkinson's disease, multiple sclerosis, and myotonic dystrophy.[62] The compound is a mild CNS stimulant that shares the pharmacological actions of amphetamines including potential for abuse.[63] Methylphenidate blocks the reuptake of dopamine and noradrenaline by presynaptic neurons.[46,64] Insomnia and nervousness are the most commonly described side effects.

Modafinil is also employed for the pharmacological treatment of EDS. This drug also blocks the reuptake of dopamine and increases the synaptic levels of dopamine.[53]

4.6.3 Serotonin

Serotonin is synthesized by neurons located in several nuclei of the midline brainstem known as raphe nuclei.[65–68] Among them, the most involved in behavioral state control are the DR and medial raphe; these nuclei are located in the mesopontine region and project directly to thalamus and cortex (Fig. 4.1). These serotonergic neurons are characterized by being active during W and turn off during REM sleep[69]; the release of serotonin as measured by microdialysis has the same profile[70] (Table 4.1). Subgroups of these serotonergic neurons have a specific activation when the experimental animal is performing stereotyped movements, such as locomotor activity or grooming.[71] Electrical stimulation of the DR produces a marked activation of the EEG.[72] In addition, photoactivation of serotonergic neurons of the DR increases W and enhances patience to wait for a delayed reward.[73,74]

Based on genetic, electrophysiological, neurochemical, and neuropharmacological approaches, it is currently accepted that serotonin is involved in the promotion of W and the inhibition of REM sleep. The serotonergic receptors can be classified into at least seven classes termed 5-HT_{1-7}.[58] The 5-HT_{1A} receptor is a somatodendritic autoreceptor, while the 5-HT_{1B} receptor is located at presynaptic sites (serotonin axon terminals). In addition, both receptor subtypes are localized at postsynaptic sites. Mutant mice that do not express 5-HT_{1A} or 5-HT_{1B} receptor exhibit greater amounts of REM sleep than their wild-type counterparts. Moreover, direct infusion of a 5-HT_{1A} receptor agonist into the DR enhances REM sleep in laboratory animals. Systemic injection of full agonists of postsynaptic 5-HT_{1A}, 5-HT_{1B}, 5-HT_{2C}, $5\text{-HT}_{2A/2C}$, 5-HT_6, and 5-HT_7 receptor increases W and reduces NREM sleep and REM sleep. Similar effects have been observed after intracerebroventricular administration of a 5-HT_3 receptor agonist.[68]

Pharmacological note. Agitation and hypervigilance accompany the serotonin syndrome, a potentially life threatening condition produced by adverse reaction or overconsumption of serotonergic drugs.[75]

Clinical studies have shown that the nonselective serotonin $5\text{-HT}_{2A/2C}$ receptor antagonists such as ritanserin, ketanserin, and seganserin increase N3 sleep in subjects with normal sleep. Ritanserin also produced an increase of N3 sleep in poor sleepers, patients with a chronic insomnia disorder, and psychiatric patients with a generalized anxiety disorder or a mood disorder. More recent evidence indicates that the selective 5-HT_{2A} receptor antagonist volinanserin, and the 5-HT_{2A} receptor inverse agonists nelotanserin and pimavanserin, significantly increase N3 sleep in subjects with normal sleep. Nelotanserin was also shown to augment N3 sleep in patients with a chronic insomnia disorder. N2 sleep tended to decrease in most of these studies, while REM sleep showed no significant changes. Thus, the coadministration of a selective 5-HT_{2A} receptor antagonist or inverse agonist along with a hypnotic drug could be a valid clinical approach for normalizing sleep induction and maintenance, as well as for promoting N3 sleep in patients with an insomnia disorder.[76]

4.6.4 Histamine

Histamine, acting via H_1 and/or H_3 receptor, has a crucial role in the regulation of the behavioral states. Histaminergic neurons are found only in the tuberomamillar nucleus of the posterior hypothalamus[77] (Fig. 4.1). These neurons also have a "W-ON" and "REM-OFF" firing profile[78] (Table 4.1). It is worth mentioning that knockout mice lacking histidine decarboxylase (enzyme involved in the synthesis of histamine) or H1 receptor are unable to stay awake when placed in a new environment.[79,80] On the contrary, drugs that increase synaptic levels of histamine augment W.[81] An optogenetic in vitro study also showed that stimulation of histaminergic neurons inhibits the NREM-promoting VLPO neurons, supporting the sleep–wake "flip-flop switch" hypothesis, and a role for histamine favoring W.[82]

Pharmacological note. Drugs that antagonize the H_1 histaminergic receptor and are prescribed as antiallergic, including diphenhydramine, hydroxyzine, and triprolidine, produce somnolence, an increased the likelihood of reduce vigilance and falling asleep.[83,84] These effects led to the use of these drugs as over-the-counter medications to promote sleep. Doxepin is the histamine H_1 receptor antagonist approved by international agencies for the treatment of an insomnia disorder. While prescribing guidelines for doxepin as an antidepressant goes as high as 300 mg daily, the approved insomnia doses are just 3 and 6 mg. This is because at low doses, this compound has minimal pharmacological activity on other neurotransmitter systems.[85]

The H_3 receptor functions as an autoreceptor, and regulates the synthesis and release of histamine. Activation of H_3 receptor reduces histamine release and promotes sleep. Conversely, blockade of H_3

receptor induces W.[86] Pitolisant, a histaminergic H_3 receptor inverse agonist, is a W-promoting drug that has been proposed for the treatment of hypersomnia.[87]

4.6.5 Acetylcholine

Cholinergic neurons involved in the maintenance of W are found in the laterodorsal and pedunculopontine tegmental nuclei (LDT-PPT), and in the BF (Fig. 4.1).[88–90] While LDT-PPT neurons project into the thalamus, the BF neurons project mainly to the cerebral cortex.

During W, and in close relationship with cortical activation, there is an increase in cholinergic neuronal firing as well as an increment in the release of the acetylcholine at cortical levels.[91–93] These neurons suppress their discharge during NREM sleep and are reactivated during REM sleep (Table 4.1).

Drugs that increase synaptic levels of acetylcholine, such as physostigmine, are capable of reversing the state of general anesthesia produced by sevoflurane in humans.[94] Recent optogenetic and "designer receptors exclusively activated by designer drugs" (DREADD) approaches have shown that direct activation of the cholinergic neurons of the BF tend to reduce slow EEG activity during NREM sleep, while inhibition of them increases the slow EEG rhythms but without an important decrease in the amount of W.[95–100]

It is considered that cholinergic neurons of the BF regulate sensory information processing, attention, and memory during W. It is important to note that the cognitive disorders characteristic of Alzheimer's disease are related, in part, to the degeneration and loss of this neuronal group.[101]

Cholinergic neurons of the medial septum and the vertical limb of the diagonal band (components of the BF) contribute to the generation of theta activity during active W (as well as during REM sleep).

Cholinergic neurons of the LDT-PPT are active mainly during active W,[91] and selective activation of the cholinergic neurons of the PPT by a chemogenetic approach suppresses EEG slow activity during NREM sleep.[102]

Pharmacological note. Cholinergic muscarinic receptor antagonist (such as atropine and scopolamine) display a dissociated state characterized by behavioral W, associated with high amplitude slow oscillations and spindles in the EEG, similar to those that occur during NREM sleep.[103] The most outstanding example of this cognitive dysfunction is the criminal use of anticholinergic substances that are present in Datura plant extracts (called "burundanga"), to induce amnesia and submissive behavior or "obedience" in victims.[104,105]

4.6.6 Hypocretins (orexins)

The hypocretins 1 and 2 (also called orexins A and B) are neuropeptides used as nonclassical neurotransmitters or neuromodulators by neurons located in the posterior and lateral hypothalamic region[106,107] (Fig. 4.1). Hypocretins act on two types of metabotropic receptors, exerting postsynaptic and presynaptic excitatory effects.[108]

The intracerebral or intraventricular administration of hypocretins facilitates the generation of W.[109] Furthermore, several experimental approaches have shown that these neurons are activated mainly during W when there is motivation. For example, experimental studies using Fos protein as an index of neuronal activity have shown that the hypocretinergic neuronal activity increases during W with motor activity related to the motivation to explore a new environment, but not during quiet wakefulness or forced locomotion.[110–112] Opto- and chemogenetic activation of these neurons also promotes W.[113–115]

This system is of great clinical importance, since the degeneration of hypocretinergic neurons is the pathogenic basis of narcolepsy with cataplexy.[7]

Pharmacological note. Intranasal hypocretin administration restored the olfactory dysfunction in patients with narcolepsy with cataplexy.[116] In addition, dual orexin receptor antagonist (DORA), suvorexant, is currently approved for the treatment of insomnia in the United States and Japan.[117]

4.6.7 Glutamate

Glutamate is the most ubiquitous neurotransmitter in the CNS. Several glutamatergic neurons located in different areas play an important role in the control of the sleep–wake cycle. With respect to W, glutamatergic neurons in the mesopontine nuclei form part of the activating systems. Some of these glutamatergic neurons are activated during W and REM sleep, while others become functionally active only during W.[91] In this regard, chemogenetic activation of glutamatergic neurons of the PPT increases W.[102] Glutamatergic neurons of the nearby medial parabrachial area are also involved in the generation of this state.[40,118] Furthermore, chemogenetic activation of glutamate-releasing neurons (Vglut2) in the supramammillary nucleus of the caudal hypothalamus produces sustained EEG activation and behavioral arousal, an effect that is virtually suppressed by genetic disruption of glutamate release from these neurons.[119] Studies in rodents have also pointed out that glutamatergic neurons of the BF also play a role in W.[97,99]

Pharmacological note. Ketamine, an NMDA receptor antagonist, inhibits W and produces sedation, hypnosis, or pharmacological coma.[120]

4.6.8 GABA

GABAergic neurons are widely distributed in the CNS.[121] Classically it was considered that GABA was a NREM sleep promoting substance. However, it is now accepted that different groups of GABAergic neurons play a variety of roles in the control of behavioral states.[122]

Experimental evidence suggests that an increase in synaptic GABA release in the nucleus pontis oralis (NPO, also called sublaterodorsal nucleus, which is the nomenclature currently more utilized in rodents), the executive area for REM sleep generation, is necessary to induce W.[123] In addition, the application of GABAergic receptor agonists into the NPO generates W.[124] Furthermore, local increase of GABA levels in the NPO prolongs the time necessary to induce general anesthesia, while isoflurane anesthesia reduces GABA levels within the NPO.[125] A group of GABAergic neurons of the BF is also involved in the generation of W; in fact, opto- and chemoactivation of GABAergic neurons that colocalize parvalbumin increase the W and fast EEG activity (gamma band, 30–60 Hz) that characterize W.[97,99,126,127]

W-promoting GABAergic neurons have also been found in the lateral hypothalamus; these neurons may induce W by inhibiting the NREM-promoting VLPO neurons, or the reticular thalamic nucleus that is critical for the generation of sleep spindles.[128,129]

GABAergic neurons of the VTA also increase their firing rate during arousal (and REM sleep)[130]; these neurons are probably involved in reward-related arousal.[122]

4.7 NEURAL SYSTEMS THAT GENERATE NREM SLEEP

Since Von Economo's early studies on the brains of patients who had died of lethargic encephalitis, the anterior region of the hypothalamus has been recognized as a critical area for sleep generation.[131] However, as reviewed below, other neuronal networks also seem to play a role in NREM sleep generation.

4.7.1 Preoptic Area

Neurons from the preoptic area (POA) of the hypothalamus are critical in the generation and maintenance of NREM sleep[132–135] (Fig. 4.1). These neurons increase their firing rate during NREM sleep, and have been identified mainly in the median POA and VLPO (Table 4.1). Most of these neurons are GABAergic, and also colocalize with the neuropeptide galanin.[136] These cells project in monosynaptic form toward the activating nuclei; in fact, electrical stimulation of the POA and adjacent BF inhibits the W-promoting neurons and induces NREM sleep.[137] Moreover, optogenetic activation or inhibition of the GABAergic VLPO neurons that project toward the W-promoting histaminergic neurons, enhances or decreases sleep, respectively.[138] Also, optogenetic stimulation of GABAergic neurons of the POA directly inhibits the hypocretinergic neurons.[139] On the other hand, experimental evidence suggests that W-promoting neurons inhibit VLPO neurons.[82,140] This reciprocal inhibition between activating and hypnogenic neurons is critical for the transition between sleep and W, and the basis of the flip-flop state switch model[19] (Fig. 4.2).

4.7.2 Medullary Reticular Formation

Early transection studies have suggested that the caudal brainstem may promote NREM sleep.[141] A recent study has identified the parafacial zone (located ventral and lateral of the genu of the facial nerve) in the medullary reticular formation as a critical area for the generation of NREM sleep[142] (Fig. 4.1). GABAergic and glycinergic neurons within this area are active during NREM sleep, and activation or inhibition of these neurons promote or suppress NREM sleep, respectively.[142,143] Furthermore, GABAergic neurons within this area directly inhibit the W-promoting medial parabrachial area. However, Sakai (2017) has shown that none of these neurons discharge maximally during NREM sleep[144]; hence, this finding does not support a role of these neurons in the control of NREM sleep. New studies are needed to reach a definite conclusion about these neurons.

4.7.3 Other NREM Sleep Promoting Areas

Neurons in the lateral habenula project to several arousal-regulating nodes. In a recent study, Gelegen and colleagues used an elegant tissue-specific genetic approach to probe the role of glutamatergic neurons in the lateral habenula, in the regulation of sleep and anesthetic states.[145] Blocking glutamatergic output from the lateral habenula fragmented NREM sleep, increased the resistance to the sedative actions of propofol, and blocked the increase in EEG power during propofol sedation. This evidence suggests that glutamatergic neurons in the lateral habenula play a role in the sedative actions of propofol, and are necessary for the consolidation of NREM sleep.

4.7.4 Critical Role of GABA in the Generation of NREM Sleep

GABAergic neurons and receptors are distributed within the activating and NREM sleep promoting system. In addition, these neurons are located in the reticular nucleus of the thalamus, where sleep spindles are generated, as well in the cortex, where they play a critical role in the generation of EEG rhythms.[146] We mentioned above that several of these GABAergic networks play an important role in the inhibition of the activating system and, as a result, the occurrence of NREM sleep.

Pharmacological note. The importance of GABA in NREM sleep generation is highlighted with the fact that benzodiazepine and nonbenzodiazepine receptor allosteric modulators (Z-drugs; zolpidem, eszopiclone, zaleplon), which are used as hypnotics, potentiate the action of GABA by acting on GABA-A receptors.[147] Also, most of the general anesthetics (such as barbiturate, etomidate, and propofol) suppress W by enhancing chloride conductance at GABA-A receptors.[148] These drugs would promote NREM sleep, sedation, or anesthesia, at least in part, by reducing the activity of the activating systems, reproducing the effect obtained by the experimental activation of the GABAergic neurons of the POA and the parafacial zone. In addition, these drugs enhance the activity of GABAergic POA neurons that inhibit the activating systems.[148]

4.8 NEURAL SYSTEMS THAT PROMOTE THE GENERATION OF REM SLEEP

4.8.1 Mesopontine Reticular Formation

The neural networks necessary and sufficient for the generation and maintenance of REM sleep are found in the mesopontine reticular formation (Fig. 4.1).[149] In fact, most of the mesopontine neurons that play a role in the maintenance of W coincide with the neurons that are responsible for the generation of REM sleep.

How does this mesopontine neuronal network generate REM sleep? Two main structural models dominate the scene. The cholinergic-aminergic model has been built from data that were originally obtained mostly from cats.[150] The other, the GABA-glutamatergic model, has been constructed from data gathered mainly from rodents.[151]

Cholinergic-aminergic model. McCarley and Hobson (1975) constructed a structural and mathematical model in which an interaction between "REM-ON" and "REM-OFF" neurons determined the REM-NREM ultradian cycle.[152] A newer version of this model, with new neuronal actors, has been also proposed.[150]

Mesopontine noradrenergic and serotonergic neurons (of the LC and DR, respectively) that are active during W turn off their activity during REM sleep ("REM-OFF" neurons) (Table 4.1). These neurons are considered "permissive" for the generation of REM sleep; that is, to generate REM sleep these neurons must be inhibited. During W, these neurons inhibit the "REM-ON" neurons of the LDT-PPT and NPO.[153–155]

Cholinergic neurons of the LDT-PPT not only increase their firing rate during W, but also do so during REM sleep[91] (Table 4.1). Optogenetic stimulation of LDT-PPT cholinergic neurons promotes transition from NREM to REM sleep.[156] Cholinergic "REM-ON" neurons of the LDT-PPT project to the NPO, the REM sleep "executive area." The release of acetylcholine within this area promotes REM sleep.[157,158]

Cholinergic neurons activate glutamatergic neurons of the NPO, which are operational during REM sleep and seem to coordinate different aspects of this state.[157,159,160] For example, these neurons may activate glycinergic neurons of the medullary reticular formation that generate the muscle atonia or, via ascending projections, contribute to the thalamocortical activation that characterize REM sleep.

New versions of the model consider that mesopontine GABAergic neurons may contribute to the inhibition of the monoaminergic neurons during REM sleep, and/or to inhibit the "REM-ON" neurons of the NPO during W.[124,150,161–167]

GABAergic-glutamatergic model. This model emphasizes the connections between "REM-OFF" GABAergic neurons of the ventrolateral periaqueductal gray (vlPAG) that modulates the "REM-ON" glutamatergic neurons of the NPO (or sublaterodorsal nucleus). Hence, GABAergic vlPAG neurons are active during W and project to the NPO.[55,151,161] In fact, bilateral inhibition of the vlPAG and adjacent areas promotes REM sleep.[166] It has been hypothesized that inhibition of vlPAG neurons during the transition from NREM to REM sleep disinhibits "REM-ON" glutamatergic neurons in the NPO triggering REM sleep. Consistent with this hypothesis, calcium imaging and unit recordings across sleep–wake states demonstrated that the activity of most GABAergic neurons in the vlPAG is strongly suppressed during REM sleep.[168] Of note is that this model, guided mostly by experiments in rodents, considers cholinergic influence on REM sleep not as determinant, but only regulatory.[169]

Pharmacological note. Most serotonin or noradrenaline reuptake inhibitors used as antidepressants tend to suppress REM sleep.[6] The inhibition of REM sleep by these drugs is used to block cataplexy (produced by the same mechanisms responsible for the REM sleep atonia) in narcoleptic patients.[64]

In humans, increasing synaptic levels of acetylcholine by acetylcholinesterase inhibitors produces W and cortical activation, while REM sleep precipitates if this drug is applied during NREM sleep.[170] These data support a bimodal role of cholinergic neurons, promoting both the generation of W and REM sleep.

4.8.2 Medullary Reticular Formation

The medullary reticular formation plays a critical role in the generation of REM sleep atonia. One accepted model considers that glutamatergic neurons in the NPO generate motor atonia by activation of glycinergic neurons in the ventromedial medulla. These neurons produce a glycinergic postsynaptic inhibition onto alpha motoneurons.[171] GABAergic neurons within this area may also play a role in the generation of REM sleep atonia.[161,172]

GABAergic neurons of the dorsal and lateral paragigantocellular nuclei inhibit "REM-OFF" neurons of the mesopontine region, and have been proposed to be also involved in the promotion of REM sleep.[173,174]

4.8.3 Other Areas Involved in REM Sleep Generation

As mentioned earlier, dopaminergic neurons of the VTA change their firing profile (from tonic to burst discharge type) and increase the release of dopamine in the nucleus accumbens and prefrontal cortex during REM sleep. These neurons may contribute to the EEG activation. In addition, ascending projections from the VTA have been suggested to be critical for the generation of dreams.[175]

The hypothalamus also participates in the control of REM sleep. Histaminergic neurons are "REM-OFF" and are probably also permissive for REM sleep generation. Furthermore, Lu and coworkers have identified a region of the POA that is active during REM sleep, the "extended" VLPO, and may promote this state.[176] However this REM sleep promoting area was not observed in the cat.[133]

Chemogenetic activation of hypocretinergic neurons suppresses REM sleep.[115] Moreover, hypocretinergic neurons decrease their firing rate during "tonic" REM sleep[177,178]; however, Fos protein and microdialysis studies conducted in cats suggest that these neurons may be active during the "phasic" components of REM sleep, and may contribute to the induction of twitches, rapid eye movements, and autonomic instability that characterize this behavioral state.[179] The role of the hypothalamic melanin-concentrating hormone (MCH) will be described in the next section.

Cholinergic and noncholinergic BF neurons are not only active during W but also during REM sleep. The latter may contribute to the cortical activation that characterizes REM sleep.[92,99,100]

Pharmacological note. Dopamine agonists induce vivid dreams, while antipsychotic drugs, which block the action of dopamine, suppress dream experiences.[175,180,181]

4.9 THE MELANIN-CONCENTRATING HORMONE: A NREM AND REM SLEEP PROMOTING SYSTEM

Neurons in the posterolateral hypothalamus and incertohypothalamic area utilize the neuropeptide MCH as a neuromodulator.[182–185] These neurons project throughout the CNS, including toward major W-promoting nuclei[185,186] (Fig. 4.1). MCH exerts its biological functions through two metabotropic receptors but only one is active in rodents.[187]

These neurons fire scarcely during W, increase their firing rate during NREM sleep, and reach a maximum during REM sleep.[188] Optogenetic and chemogenetic studies strongly suggest that this system plays a critical role in the generation of sleep.[189,190] Since the administration of MCH into the cerebral ventricles, POA, BF, DR nucleus, LC, and NPO facilitates the generation of NREM sleep and/or REM sleep, it is possible that MCHergic neurons inhibit the activity of the activating systems and/or activate the hypnogenic nuclei to promote sleep.[191–196] For example, MCHergic neurons inhibit DR serotonergic neurons and decrease serotonergic release[197,198]; this finding may explain, at least in part, the promotion of REM sleep induced by MCH.[191] The suppression of the serotonergic activity by MCH could explain also the pro-depressive effect of this neuropeptide.[199–203]

A recent chemogenetic study suggests that when MCHergic neurons are physiologically recruited, NREM sleep depth is increased and the extinction of NREM sleep episodes is accelerated, strengthening the probability for natural NREM to REM sleep transition.[204]

On the other hand, W-promoting neurotransmitter/neuromodulators tend to silence the MCHergic neurons.[183]

Pharmacological note. Experimental evidence strongly suggests that MCH antagonists may be a therapeutic approach to treat depression and anxiety.[205]

4.10 OTHER SUBSTANCES THAT PARTICIPATE IN THE CONTROL OF WAKEFULNESS AND SLEEP

4.10.1 Melatonin

Melatonin is a hormone that is secreted during the night from the pineal gland.[206] As mentioned, commands from SCN promote sleep during the night (and/or W during daytime, process C) in diurnal animals. This nucleus, in addition to modulating W and sleep promoting areas, regulates the release of melatonin, through indirect modulation of the sympathetic system. Hence, melatonin is a hormone that signals the absence of light, and prepares the individual physiology to the night period.[207] In humans, melatonin has a weak sleep-promoting effect; recently, an inhibitory effect of melatonin onto the hypocretinergic neurons has been described.[208]

Pharmacological note. Because melatonin is the physiological marker of the night, it is usually indicated for circadian maladjustments such as jet lag. Also, melatonin and melatonin receptor agonists including ramelteon and tasimelteon are used to treat certain types of insomnia.[147] In this respect, ramelteon has a specific indication for difficulty with sleep onset, while tasimelteon is indicated for the treatment on non-24-hour circadian rhythm sleep–wake disorder. Finally, agomelatine, marketed as an antidepressant, has multiple receptor effects that include melatonin receptor agonist activity.[85]

4.10.2 Adenosine

Process S determines an increase of sleep pressure during a prolonged W period. Different lines of research have shown that during W, NREM sleep promoting substances such as adenosine are released into the synaptic space, at least in part, from astrocytes (gliotransmission).[25,26,209] Adenosine promotes sleep by inhibiting the activating systems and stimulating the NREM-promoting neurons of the POA.[26,210–212]

Recent experimental evidence suggested that GABAergic neurons of the nucleus accumbens (ventral striatum) that express the adenosinergic A2a receptor and dopaminergic receptors are critical for sleep and W regulation.[213,214] In fact, opto- or chemogenetic activation of these neurons induces NREM sleep.[215]

Pharmacological note. Caffeine, the world's most consumed psychostimulant, promotes W by blocking adenosine receptors.[216] Caffeine is usually obtained from different sources such as coffee, cola drinks, and caffeine pills.[217] In addition, several natural products that are consumed as tea have an important amount of caffeine. An example is the W-promoting *Ilex paraguariensis*, known as yerba mate, a traditional beverage in the south of Latin America that has now expanded to other world regions.[218,219]

4.10.3 Cytokines

It has been shown that loss of sleep distorts immune function. In fact, messenger molecules of immune function (cytokines) are present in the normal brain and interact with the neurochemical systems that regulate the sleep–wake cycle.[220–222] The cytokines interleukin-1 (IL-1) and tumor necrosis factor (TNF) promote NREM sleep, increasing the activity of the NREM-promoting neurons of the POA. Part of this effect seems to be mediated by prostaglandin D2, which is another NREM-promoting factor. The role of cytokines in the generation of NREM sleep could occur under normal conditions, but it is potentiated in pathological conditions where the synthesis of these cytokines increase; for example, IL-1 and TNF would be involved with the hypersomnia that accompanies infection and fever.

4.10.4 Endocannabinoids

The predominant transmitters for the endocannabinoid system are anandamide (AEA) and 2-archidonylglycerol (2-AG). These molecules are ubiquitous in the CNS and are released during periods of neuronal activity.[223] These neuromodulators produce the majority of their central effects by activating the cannabinoid receptor 1 (CB1); activation of this G-protein-coupled receptor reduces neurotransmitter release at many synapses. Administration of exogenous AEA consistently increases REM and NREM sleep. However, conflicting results arise from attempts to increase endogenous AEA levels. A recent study suggests that endocannabinoid signaling through CB1 is necessary for NREM stability but it is not necessary for sleep homeostasis.[224]

Pharmacological note. Since antiquity, cannabinoids have been used as a treatment for insomnia[225]; however, there is not enough experimental and clinical evidence yet to support the use of cannabis for this condition.

4.10.5 Nitric Oxide

Several groups of neurons produce nitric oxide (NO); among them, the cholinergic neurons of the LDT-PPT that play a critical role both in W and REM sleep.[226]

Mariño and Cudeiro (2003) suggested an activating role of NO.[227] However, the intracerebroventricular injection of the precursor of NO, L-arginine, produces an increase in NREM sleep in rats.[228] Similar effects have been seen with the use of NO donors in rats and cats.[229–231] Other studies have shown that inhibitors of NO synthase (which decreases the release of NO) produce a decrease of sleep,[228,229,232] and attenuate the homeostatic "rebound" of sleep that follows its deprivation.[233] Neurons that contain NO synthase (and colocalize with GABA) in the cerebral cortex are active during

NREM sleep.[234] These neurons are thought to synchronize slow cortical electrical rhythms during NREM sleep, probably in relation with homeostatic sleep drive.[235,236] Finally, the release of NO increases during W or sleep according to the region where it is studied.[237]

In summary, available data suggest that NO promotes W or sleep depending on the neuronal network under study.

4.10.6 Other Substances

There is experimental evidence that substances such as various neuropeptides (neuropeptide Y, neuropeptide-VF, nesfatin-1, etc.)[238–240] and hormones (GHRH, female reproductive hormones, etc.)[241,242] also have a regulatory role in the sleep–wake cycle. For example, Jego et al. (2012) have shown that the neuropeptide nesfatin-1 is coexpressed with MCH in the lateral hypothalamic area.[243] These authors also showed that disruption of the brain nesfatin-1 signaling suppressed REM sleep with only small alteration of NREM sleep.

4.11 PHARMACOLOGICAL APPROACHES IN SLEEP PATHOLOGY

Knowing the neurobiological basis of W and sleep provides the clinician with the frame to understand sleep pathologies and the pharmacological approaches for their treatment. Below we discuss some of the neurochemical rationale for the therapeutic approach for paradigmatic sleep conditions.

4.11.1 Insomnia

An inability to fall or stay asleep, and daytime complaints such as somnolence and fatigue that occurs no less than three times per week, for at least 3 months, characterize the chronic insomnia disorder in adults.[1]

An imbalance between the W and NREM sleep promoting mechanisms is likely behind this disorder. Behavioral approaches and sleep hygiene may be enough to balance this disorder. If pharmacological treatment is needed, the medications approved for the treatment of this disorder include benzodiazepine (BZD) receptor allosteric modulators, either BZD (triazolam, temazepam, flurazepam) or non-BZD (zolpidem, eszopiclone, zaleplon) agents. These drugs promote sleep, at least in part, by reducing the activity of the activating systems. The melatonin receptor agonist ramelteon and low-dose doxepin, whose mechanism of action is mainly related to the blockade of histamine H1 receptor, are also employed for the treatment of insomnia. These drugs improve sleep onset latency (zolpidem immediate-release, zaleplon, ramelteon) and/or sleep maintenance (temazepam, flurazepam, zolpidem extended-release, eszopiclone, low-dose doxepin). However, during their administration, N3 sleep and REM sleep do not regain normal levels or can be even further reduced.[76,147]

A new pharmacological approach for the treatment of an insomnia disorder is now available. It refers to the dual orexin (hypocretin) receptor antagonist (DORA) suvorexant that blocks the effect of the W-promoting orexin. In this respect, suvorexant administration was associated with significant improvements in time to sleep onset, total sleep time, and subjective quality of sleep in young and middle-aged patients with chronic insomnia. The greater amounts of total sleep time were related to an increase in REM sleep and N2 sleep.[117,147]

4.11.2 Hypersomnia

Hypersomnia disorder is a term used for a group of disorders in which the primary characteristic is EDS in the presence of normal or longer than normal nocturnal sleep.[244] These disorders are currently classified as central disorders of hypersomnolence.[1] Among these disorders, the most common is narcolepsy. The management of this pathology includes several behavioral approaches and pharmacological treatment. The most common symptom is EDS, and the first pharmacological approach is to use modafinil. This drug blocks the DAT, increasing dopamine synaptic levels. As mentioned before, increasing dopamine levels has a wake-promoting effect. Of note is that genetic ablation of the DAT abolishes the wake-promoting effect of modafinil. However, there are other possible sites of action of this drug.[245] Amphetamine-like drugs such as methylphenidate, that also increase the synaptic monoamines, are used, in addition, to reduce sleepiness.[246]

A new therapeutic approach is the administration of pitolisant, an inverse agonist/antagonist of the histamine H3 receptor that is an inhibitory autoreceptor. As a result, pitolisant increases the histaminergic tone, promoting W. Pitolisant decreases EDS and cataplexy rate in narcoleptic patients.[247,248]

4.11.3 Parasomnias

Parasomnias are unpleasant or undesirable behavioral phenomena that occur during the sleep period. There are different types of parasomnias, that are classified in REM and NREM sleep parasomnias.[1,249] One of them is the REM sleep behavior disorder (RBD). During RBD the REM sleep atonia does not occur, and the patients act out their dreams. Severe injuries can occur during the REM without atonia episodes. About 90% of patients with chronic RBD respond well to clonazepam (0.5–2 mg) administered half an hour before sleep time.[250] Clonazepam is a benzodiazepine whose mechanism of action for the RBD is still unknown. It has been tentatively proposed that the improvement of sleep would depend upon a reduction of REM sleep time and/or intensity.

4.11.4 Circadian Disorders of the Sleep–Wake Cycle

Delayed and advanced sleep phase disorders are the most common circadian conditions. These are characterized by bedtimes and wake-up times that are delayed or advanced 3 or more hours, respectively. The free-running disorder occurs predominantly in severely blind individuals and is characterized by steady daily delaying drift of the major sleep period.

The goal of the therapy is to align the timing of the circadian clock (the SCN) with the desired light-night cycle. Sleep hygiene and chronotherapy with light and/or melatonin (or melatonin agonists) are usually indicated.[251]

4.11.5 Jet Lag and Shift Work Disorder

Jet lag and shift work disorder are circadian rhythm sleep–wake disorders resulting from altered sleep–wake schedule in relation to the external environment. As for the circadian disorders, the treatment aim is also to realign the internal circadian clock with the external environment. Behavioral therapies such as sleep hygiene and management of the light-dark and sleep schedule are indicated. Pharmacologic agents are used to treat insomnia and excessive sleepiness, and melatonin (or melatonin agonists) is used to facilitate sleep and circadian realignment.[252]

4.11.6 Neurological and Psychiatric Disorders That Affect Sleep

Several neurological and psychiatric and general medical disorders may affect sleep. Paradigmatic examples of a psychiatric and a neurological disorder that affect sleep are depression and Parkinson's disease.

Depression is associated with disturbed sleep, including disturbances of sleep continuity, reduced N3 sleep, and altered REM sleep parameters. Activating antidepressants may worsen sleep, which is related to their mechanism of action including inhibition of serotonin and/or noradrenaline reuptake, and activation of serotonin 5-HT_{1A} and 5-HT_2 and noradrenaline α_1 receptors. Worsening of sleep in patients with a diagnosis of depression has been described following treatment with the activating tricyclic antidepressants imipramine, desipramine, and protriptyline; the selective serotonin reuptake inhibitors fluoxetine and paroxetine; the selective noradrenaline reuptake inhibitor reboxetine; and the serotonin and noradrenaline reuptake inhibitors venlafaxine, duloxetine, and milnacipran. Sedative antidepressants include the tricyclic derivatives amitriptyline, nortriptyline, trimipramine, and doxepin, and the serotonin reuptake inhibitors trazodone and nefazodone; their actions on sleep variables depend upon the blockade of serotonin 5-HT_{2A} and histamine H_1 receptors.[6,253–255]

Parkinson's disease is a neurological pathology that courses with degeneration of dopaminergic neurons.[256] Insomnia is the most frequent sleep disorder in Parkinson's disease, and is linked to motor symptoms that characterize the pathology. The sleep disorder is mainly characterized by increased sleep latency and intra-sleep awakenings. The administration of L-Dopa (administered together with the dopamine decarboxylase inhibitor carbidopa) may improve sleep-associated motor symptoms that contribute to insomnia. The dopamine agonists ropinirole, pramipexole, rotigotine patch, and the monoamine oxidase B inhibitor rasagiline have also been shown to recover insomnia in Parkinson's disease patients.[257]

4.12 CONCLUSIONS

A detailed knowledge of the neurochemistry of the activating and hypnogenic systems is important to understand and treat sleep pathologies. An exemplary achievement in relation to the activating systems has been the unveiling of the pathogenesis of narcolepsy. This pathology is caused by the degeneration of hypocretinergic neurons,[7] which prompted paraclinical studies such as the titration of hypocretin-1 in the cerebrospinal fluid for diagnostic confirmation of narcolepsy, and therapeutic advances such as intranasal hypocretin-1 administration for the treatment of some aspects of the disease,[116] as well as hypocretin antagonist for the treatment of insomnia.[258]

Acknowledgments

This study was supported by the "Agencia Nacional de Investigación e Innovación, Fondo Clemente Estable FCE-1-2017-1-136550" grant, the "Comisión Sectorial de Investigación Científica I + D-2016-589" grant, and the "Programa de Desarrollo de Ciencias Básicas, PEDECIBA" from Uruguay.

References

1. American-Academy-of-Sleep-Medicine. *International classification of sleep disorders: diagnostic and coding manual*. 3rd ed. Darien, IL: American Academy of Sleep Medicine; 2014.
2. Carskadon MA, Dement W. Normal human sleep: an overview. In: Kryger MH, et al., editors. *Principles and practices of sleep medicine*. Philadelphia, PA: Elsevier-Saunders; 2011. p. 16–26.
3. Pace-Schott E. The neurobiology of dreaming. In: Kryger MH, et al., editors. *Principles and practices of sleep medicine*. Philadelphia, PA: Elsevier-Saunders; 2011. p. 563–75.
4. Dement W, Kleitman N. The relation of eye movements during sleep to dream activity: an objective method for the study of dreaming. *J Exp Psychol* 1957;**53**(5):339–46.
5. Siclari F, et al. The neural correlates of dreaming. *Nat Neurosci* 2017;**20**(6):872–8.
6. Palagini L, Baglioni C, Ciapparelli A, Gemignani A, Riemann D. REM sleep dysregulation in depression: state of the art. *Sleep Med Rev* 2013;**17**:377–90.
7. Mignot E. Narcolepsy: pathophysiology and genetic predisposition. In: Krieger MH, et al., editors. *Principles and practices of sleep medicine*. Philadelphia, PA: Saunders; 2011. p. 938–56.
8. Peirano PD, Algarin CR. Sleep in brain development. *Biol Res* 2007;**40**(4):471–8.
9. Roffwarg HP, Muzio JN, Dement WC. Ontogenetic development of the human sleep-dream cycle. *Science* 1966;**152**(3722):604–19.
10. Hobson JA. Dreaming as delirium: a mental status analysis of our nightly madness. *Semin Neurol* 1997;**17**(2):121–8.
11. Steriade M, McCormick DA, Sejnowski TJ. Thalamocortical oscillations in the sleeping and aroused brain. *Science* 1993;**262**(5134):679–85.
12. Crunelli V, David F, Lorincz ML, Hughes SW. The thalamocortical network as a single slow wave-generating unit. *Curr Opin Neurobiol* 2015;**31**:72–80.
13. Huguenard JR, McCormick DA. Thalamic synchrony and dynamic regulation of global forebrain oscillations. *Trends Neurosci* 2007;**30**(7):350–6.
14. Fuentealba P, Steriade M. The reticular nucleus revisited: intrinsic and network properties of a thalamic pacemaker. *Prog Neurobiol* 2005;**75**(2):125–41.
15. Montagna P. Fatal familial insomnia: a model disease in sleep physiopathology. *Sleep Med Rev* 2005;**9**(5):339–53.
16. Heinzer R, Series F. Normal physiology of the upper and lower airways. In: Kryger MH, et al., editors. *Principles and practices of sleep medicine*. Philadelphia, PA: Elsevier-Saunders; 2011. p. 259–68.
17. Amici R, Cerri M, Parmeggiani PL. Overview of physiological processes during sleep. In: Kushida C, editor. *The Encyclopedia of Sleep*, vol. 1, Waltham, MA: Academic Press; 2013. p. 385–9.
18. Brando V, Castro-Zaballa S, Falconi A, Torterolo P, Migliaro ER. Statistical, spectral and non-linear analysis of the heart rate variability during wakefulness and sleep. *Arch Ital Biol* 2014;**152**(1):32–46.
19. Saper CB, Fuller PM, Pedersen NP, Lu J, Scammell TE. Sleep state switching. *Neuron* 2010;**68**(6):1023–42.
20. Borbely AA. A two process model of sleep regulation. *Hum Neurobiol* 1982;**1**(3):195–204.
21. Mistlberger RE. Circadian regulation of sleep in mammals: role of the suprachiasmatic nucleus. *Brain Res Brain Res Rev* 2005;**49**(3):429–54.
22. Welsh DK, Takahashi JS, Kay SA. Suprachiasmatic nucleus: cell autonomy and network properties. *Annu Rev Physiol* 2010;72551–77.
23. Vujovic N, Gooley JJ, Jhou TC, Saper CB. Projections from the subparaventricular zone define four channels of output from the circadian timing system. *J Comp Neurol* 2015;**523**(18):2714–37.

24. Chou TC, et al. Critical role of dorsomedial hypothalamic nucleus in a wide range of behavioral circadian rhythms. *J Neurosci* 2003;**23**(33):10691–702.
25. Basheer R, Strecker RE, Thakkar MM, McCarley RW. Adenosine and sleep-wake regulation. *Prog Neurobiol* 2004;**73**(6):379–96.
26. Huang ZL, Urade Y, Hayaishi O. The role of adenosine in the regulation of sleep. *Curr Top Med Chem* 2011;**11**(8):1047–57.
27. Roky R, Kapas L, Taishi TP, Fang J, Krueger JM. Food restriction alters the diurnal distribution of sleep in rats. *Physiol Behav* 1999;**67**(5):697–703.
28. Jenkins JB, et al. Sleep is increased in mice with obesity induced by high-fat food. *Physiol Behav* 2006;**87**(2):255–62.
29. Varin C, et al. Glucose induces slow-wave sleep by exciting the sleep-promoting neurons in the ventrolateral preoptic nucleus: a new link between sleep and metabolism. *J Neurosci* 2015;**35**(27):9900–11.
30. Burdakov D, Gerasimenko O, Verkhratsky A. Physiological changes in glucose differentially modulate the excitability of hypothalamic melanin-concentrating hormone and orexin neurons in situ. *J Neurosci* 2005;**25**(9):2429–33.
31. Moruzzi G, Magoun HW. Brain stem reticular formation and activation of the EEG. *Electroencephalogr Clin Neurophysiol* 1949;**1**(4):455–73.
32. Velluti RA. Interactions between sleep and sensory physiology. *J Sleep Res* 1997;**6** (2):61–77.
33. Amici R, et al. Changes in REM sleep occurrence due to rhythmical auditory stimulation in the rat. *Brain Res* 2000;**868**(2):241–50.
34. Amici R, et al. REM sleep enhancement due to rhythmical auditory stimulation in the rat. *Behav Brain Res* 2001;123155–63.
35. Guillaumin MCC, et al. Cortical region-specific sleep homeostasis in mice: effects of time of day and waking experience. *Sleep* 2018.
36. Vyazovskiy VV, et al. Local sleep in awake rats. *Nature* 2011;**472**(7344):443–7.
37. Ding F, et al. Changes in the composition of brain interstitial ions control the sleep-wake cycle. *Science* 2016;**352**(6285):550–5.
38. Torterolo P, Sampogna S, Chase MH. A restricted parabrachial pontine region is active during non-rapid eye movement sleep. *Neuroscience* 2011;190184–93.
39. Torterolo P, Vanini G. New concepts in relation to generating and maintaining arousal. *Rev Neurol* 2010;**50**(12):747–58.
40. Fuller PM, Sherman D, Pedersen NP, Saper CB, Lu J. Reassessment of the structural basis of the ascending arousal system. *J Comp Neurol* 2011;**519**(5):933–56.
41. Aston-Jones G, Bloom FE. Activity of norepinephrine-containing locus coeruleus neurons in behaving rats anticipates fluctuations in the sleep-waking cycle. *J Neurosci* 1981;**1**(8):876–86.
42. Berridge CW, Abercrombie ED. Relationship between locus coeruleus discharge rates and rates of norepinephrine release within neocortex as assessed by in vivo microdialysis. *Neuroscience* 1999;**93**(4):1263–70.
43. Foote SL, Berridge CW, Adams LM, Pineda JA. Elecrophysiological evidence for the involvement of the locus coeruleus in the alerting, orienting, and attending. In: Barnes CD, Pompeiano O, editors. *Progress in brain research*. Amsterdam: Elsevier; 1991. p. 521–31.
44. Carter ME, et al. Tuning arousal with optogenetic modulation of locus coeruleus neurons. *Nat Neurosci* 2010;**13**(12):1526–33.
45. Gompf HS, et al. Locus ceruleus and anterior cingulate cortex sustain wakefulness in a novel environment. *J Neurosci* 2010;**30**(43):14543–51.
46. Nishino S, Mignot E. Pharmacological aspects of human and canine narcolepsy. *Prog Neurobiol* 1997;**52**(1):27–78.

47. Nelson LE, et al. The alpha2-adrenoceptor agonist dexmedetomidine converges on an endogenous sleep-promoting pathway to exert its sedative effects. *Anesthesiology* 2003;**98**(2):428–36.
48. Oades RD, Halliday GM. Ventral tegmental (A10) system: neurobiology. 1. Anatomy and connectivity. *Brain Res* 1987;**434**(2):117–65.
49. Schultz W, Apicella P, Ljungberg T. Responses of monkey dopamine neurons to reward and conditioned stimuli during successive steps of learning a delayed response task. *J Neurosci* 1993;**13**(3):900–13.
50. Wightman RM, Robinson DL. Transient changes in mesolimbic dopamine and their association with 'reward'. *J Neurochem* 2002;**82**(4):721–35.
51. Lena I, et al. Variations in extracellular levels of dopamine, noradrenaline, glutamate, and aspartate across the sleep–wake cycle in the medial prefrontal cortex and nucleus accumbens of freely moving rats. *J Neurosci Res* 2005;**81**(6):891–9.
52. Dahan L, et al. Prominent burst firing of dopaminergic neurons in the ventral tegmental area during paradoxical sleep. *Neuropsychopharmacology* 2007;**32**(6):1232–41.
53. Wisor JP, et al. Dopaminergic role in stimulant-induced wakefulness. *J Neurosci* 2001;**21**(5):1787–94.
54. Eban-Rothschild A, Rothschild G, Giardino WJ, Jones JR, de Lecea L. VTA dopaminergic neurons regulate ethologically relevant sleep-wake behaviors. *Nat Neurosci* 2016;**19** (10):1356–66.
55. Lu J, Sherman D, Devor M, Saper CB. A putative flip-flop switch for control of REM sleep. *Nature* 2006;**441**(7093):589–94.
56. Cho JR, et al. Dorsal raphe dopamine neurons modulate arousal and promote wakefulness by salient stimuli. *Neuron* 2017;**94**(6):1205–19.
57. Monti JM, Monti D. The involvement of dopamine in the modulation of sleep and waking. *Sleep Med Rev* 2007;**11**(2):113–33.
58. Monti JM, Jantos H. The roles of dopamine and serotonin, and of their receptors, in regulating sleep and waking. *Prog Brain Res* 2008;172625–46.
59. Buysee DJ, Schweitzer PK, Moul DE. Clinical pharmacology of other drugs used as hypnotics. In: Kryger MH, et al., editors. *Principles and practices of sleep medicine*. Philadelphia, PA: Elsevier-Saunders; 2005. p. 452–67.
60. Monti JM, Torterolo P, Pandi Perumal SR. The effects of second generation antipsychotic drugs on sleep variables in healthy subjects and patients with schizophrenia. *Sleep Med Rev* 2017;3351–7.
61. Verghese C, Abdijadid S. *Methylphenidate*. Treasure Island, FL: StatPearls; 2018.
62. Monti JM. The pharmacological treatment of sleep disorders. In: Chokroverty S, Billiard M, editors. *Sleep medicine*. New York: Springer; 2015. p. 527–32.
63. Billiard M, Lubin S. Modafinil: development and use of the compound. In: Chokroverty S, Billiard M, editors. *Sleep medicine*. New York: Springer; 2015. p. 541–4.
64. Mignot E. Narcolepsy: pharmacology, pathophysiology and genetics. In: Kryger MH, et al., editors. *Principles and practices of sleep medicine*. Philadelphia, PA: Saunders; 2005. p. 761–99.
65. Jacobs BL, Azmitia EC. Structure and function of the brain serotonin system. *Physiol Rev* 1992;**72**(1):165–229.
66. Monti JM. The role of dorsal raphe nucleus serotonergic and non-serotonergic neurons, and of their receptors, in regulating waking and rapid eye movement (REM) sleep. *Sleep Med Rev* 2010;**14**(5):319–27.
67. Monti JM. The structure of the dorsal raphe nucleus and its relevance to the regulation of sleep and wakefulness. *Sleep Med Rev* 2010;14307–17.
68. Monti JM. Serotonin control of sleep-wake behavior. *Sleep Med Rev* 2011;**15**(4):269–81.

69. McGinty DJ, Harper RM. Dorsal raphe neurons: depression of firing during sleep in cats. *Brain Res* 1976;**101**(3):569–75.
70. Portas CM, Bjorvatn B, Ursin R. Serotonin and the sleep/wake cycle: special emphasis on microdialysis studies. *Prog Neurobiol* 2000;**60**(1):13–35.
71. Jacobs BL, Fornal CA. Brain serotonergic neuronal activity in behaving cats. In: Monti JM, et al., editors. *Serotonin and sleep: molecular, functional and clinical aspects*. Basel, Boston, Berlin: Birkhauser; 2008.
72. Dringenberg HC, Vanderwolf CH. Neocortical activation: modulation by multiple pathways acting on central cholinergic and serotonergic systems. *Exp Brain Res* 1997;**116**(1):160–74.
73. Ito H, et al. Analysis of sleep disorders under pain using an optogenetic tool: possible involvement of the activation of dorsal raphe nucleus-serotonergic neurons. *Mol Brain* 2013;**6**:59.
74. Miyazaki KW, et al. Optogenetic activation of dorsal raphe serotonin neurons enhances patience for future rewards. *Curr Biol* 2014;**24**(17):2033–40.
75. Boyer EW, Shannon M. The serotonin syndrome. *N Engl J Med* 2005;**352** (11):1112–20.
76. Monti JM, Torterolo P, Spence DW, Pandi-Perumal SR. Selective serotonin 5-HT2A receptor antagonists and inverse agonists specifically promote slow wave sleep (Stage N3) in man. *Sleep Vigilance* 2018;**2**:2–23.
77. Monti JM. The role of tuberomammillary nucleus histaminergic neurons, and of their receptors, in the regulation of sleep and waking. In: Mallick BN, Pandi-Perumal SR, McCarley RW, Morrison AR, editors. *REM sleep: regulation and function*. Cambridge University Press; 2011. p. 223–33.
78. Takahashi K, Lin JS, Sakai K. Neuronal activity of histaminergic tuberomammillary neurons during wake-sleep states in the mouse. *J Neurosci* 2006;**26**(40):10292–8.
79. Parmentier R, et al. Anatomical, physiological, and pharmacological characteristics of histidine decarboxylase knock-out mice: evidence for the role of brain histamine in behavioral and sleep-wake control. *J Neurosci* 2002;**22**(17):7695–711.
80. Parmentier R, et al. Role of histamine H1-receptor on behavioral states and wake maintenance during deficiency of a brain activating system: a study using a knockout mouse model. *Neuropharmacology* 2016;10620–34.
81. Kalivas PW. Histamine-induced arousal in the conscious and pentobarbital-pretreated rat. *J Pharmacol Exp Ther* 1982;**222**(1):37–42.
82. Williams RH, et al. Optogenetic-mediated release of histamine reveals distal and autoregulatory mechanisms for controlling arousal. *J Neurosci* 2014;**34**(17):6023–9.
83. Roth T, Roehrs T, Koshorek G, Sicklesteel J, Zorick F. Sedative effects of antihistamines. *J Allergy Clin Immunol* 1987;**80**(1):94–8.
84. Monti JM. Involvement of histamine in the control of the waking state. *Life Sci* 1993;**53** (17):1331–8.
85. Neubauer DN, Pandi-Perumal SR, Spence DW, Buttoo K, Monti JM. Pharmacotherapy of insomnia. *J Cent Nerv Syst Dis* 2018;**10** 1179573518770672.
86. Thakkar MM. Histamine in the regulation of wakefulness. *Sleep Med Rev* 2011;**15** (1):65–74.
87. Leu-Semenescu S, Nittur N, Golmard JL, Arnulf I. Effects of pitolisant, a histamine H3 inverse agonist, in drug-resistant idiopathic and symptomatic hypersomnia: a chart review. *Sleep Med* 2014;**15**(6):681–7.
88. Jones B. Basic mechanisms of sleep-wake states. In: Kryger MH, et al., editors. *Principles and practices of sleep medicine*. Philadelphia, PA: Elsevier-Saunders; 2005. p. 136–53.
89. Semba K. Multiple output pathways of the basal forebrain: organization, chemical heterogeneity, and roles in vigilance. *Behav Brain Res* 2000;**115**(2):117–41.

90. Satoh K, Fibiger HC. Cholinergic neurons of the laterodorsal tegmental nucleus: efferent and afferent connections. *J Comp Neurol* 1986;**253**(3):277–302.
91. Boucetta S, Cisse Y, Mainville L, Morales M, Jones BE. Discharge profiles across the sleep-waking cycle of identified cholinergic, GABAergic, and glutamatergic neurons in the pontomesencephalic tegmentum of the rat. *J Neurosci* 2014;**34**(13):4708–27.
92. Lee MG, Hassani OK, Alonso A, Jones BE. Cholinergic basal forebrain neurons burst with theta during waking and paradoxical sleep. *J Neurosci* 2005;**25**(17):4365–9.
93. Marrosu F, et al. Microdialysis measurement of cortical and hippocampal acetylcholine release during sleep-wake cycle in freely moving cats. *Brain Res* 1995;**671** (2):329–32.
94. Plourde G, Chartrand D, Fiset P, Font S, Backman SB. Antagonism of sevoflurane anaesthesia by physostigmine: effects on the auditory steady-state response and bispectral index. *Br J Anaesth* 2003;**91**(4):583–6.
95. Chen L, et al. Basal forebrain cholinergic neurons primarily contribute to inhibition of electroencephalogram delta activity, rather than inducing behavioral wakefulness in mice. *Neuropsychopharmacology* 2016;**41**(8):2133–46.
96. Shi YF, Han Y, Su YT, Yang JH, Yu YQ. Silencing of cholinergic basal forebrain neurons using archaerhodopsin prolongs slow-wave sleep in mice. *PLoS One* 2015;**10**(7): e0130130.
97. Anaclet C, et al. Basal forebrain control of wakefulness and cortical rhythms. *Nat Commun* 2015;68744.
98. Irmak SO, de Lecea L. Basal forebrain cholinergic modulation of sleep transitions. *Sleep* 2014;**37**(12):1941–51.
99. Xu M, et al. Basal forebrain circuit for sleep-wake control. *Nat Neurosci* 2015;**18** (11):1641–7.
100. Han Y, et al. Selective activation of cholinergic basal forebrain neurons induces immediate sleep-wake transitions. *Curr Biol* 2014;**24**(6):693–8.
101. Vitiello MV, Borson S. Sleep disturbances in patients with Alzheimer's disease: epidemiology, pathophysiology and treatment. *CNS Drugs* 2001;**15**(10):777–96.
102. Kroeger D, et al. Cholinergic, glutamatergic, and GABAergic neurons of the pedunculopontine tegmental nucleus have distinct effects on sleep/wake behavior in mice. *J Neurosci* 2017;**37**(5):1352–66.
103. Sannita WG, Maggi L, Rosadini G. Effects of scopolamine (0.25-0.75 mg i.m.) on the quantitative EEG and the neuropsychological status of healthy volunteers. *Neuropsychobiology* 1987;**17**(4):199–205.
104. Ardila-Ardila A, Moreno CB, Ardila-Gomez SE. Scopolamine poisoning ('burundanga'): loss of the ability to make decisions. *Rev Neurol* 2006;**42**(2):125–8.
105. Ardila A, Moreno C. Scopolamine intoxication as a model of transient global amnesia. *Brain Cogn* 1991;**15**(2):236–45.
106. de Lecea L, et al. The hypocretins: hypothalamus-specific peptides with neuroexcitatory activity. *Proc Natl Acad Sci USA* 1998;**95**(1):322–7.
107. Sakurai T, et al. Orexins and orexin receptors: a family of hypothalamic neuropeptides and G protein-coupled receptors that regulate feeding behavior. *Cell* 1998;**92** (4):573–85.
108. Torterolo P, Monti J, Pandi-Perumal SR. Role of orexin on sleep: interactions with other neurotransmitter systems. In: Sakurai T, et al., editors. *Orexin and sleep*. Switzerland: Springer; 2015. p. 181–202.
109. Piper DC, Upton N, Smith MI, Hunter AJ. The novel brain neuropeptide, orexin-A, modulates the sleep-wake cycle of rats. *Eur J Neurosci* 2000;**12**(2):726–30.
110. Torterolo P, Yamuy J, Sampogna S, Morales FR, Chase MH. Hypocretinergic neurons are primarily involved in activation of the somatomotor system. *Sleep* 2003;125–8.

111. Torterolo P, Yamuy J, Sampogna S, Morales FR, Chase MH. Hypothalamic neurons that contain hypocretin (orexin) express c-fos during active wakefulness and carbachol-induced active sleep. *Sleep Res Online* 2001;**4**(1):25–32 http://www.sro.org/2001/Torterolo/25.
112. Torterolo P, Vanini G, Cabrera G, Chase M, Falconi A. Hypocretins (orexins) in the inferior colliculus: anatomy and physiology. *Sleep Med* 2009;**10**(Suppl. 2):S61.
113. Adamantidis AR, Zhang F, Aravanis AM, Deisseroth K, de Lecea L. Neural substrates of awakening probed with optogenetic control of hypocretin neurons. *Nature* 2007;**450**(7168):420–4.
114. Carter ME, Adamantidis A, Ohtsu H, Deisseroth K, de Lecea L. Sleep homeostasis modulates hypocretin-mediated sleep-to-wake transitions. *J Neurosci* 2009;**29**(35):10939–49.
115. Sasaki K, et al. Pharmacogenetic modulation of orexin neurons alters sleep/wakefulness states in mice. *PLoS One* 2011;**6**(5):e20360.
116. Baier PC, et al. Olfactory dysfunction in patients with narcolepsy with cataplexy is restored by intranasal Orexin A (Hypocretin-1). *Brain* 2008;**131**(Pt 10):2734–41.
117. Kuriyama A, Tabata H. Suvorexant for the treatment of primary insomnia: a systematic review and meta-analysis. *Sleep Med Rev* 2017;351–7.
118. Kaur S, et al. Glutamatergic signaling from the parabrachial nucleus plays a critical role in hypercapnic arousal. *J Neurosci* 2013;**33**(18):7627–40.
119. Pedersen NP, et al. Supramammillary glutamate neurons are a key node of the arousal system. *Nat Commun* 2017;**8**(1):1405.
120. Wolff K, Winstock AR. Ketamine: from medicine to misuse. *CNS Drugs* 2006;**20**(3):199–218.
121. Mugnaini E, Oertel WH. An atlas of the distribution of GABAergic neurons and terminals. In: Bjorklund A, Hokfelt T, editors. *Handbook of chemical neuroanatomy*. Amsterdam: Elsevier; 1985. p. 436–608.
122. Brown RE, McKenna JT. Turning a negative into a positive: ascending GABAergic control of cortical activation and arousal. *Front Neurol* 2015;6135.
123. Vanini G, Wathen BL, Lydic R, Baghdoyan HA. Endogenous GABA levels in the pontine reticular formation are greater during wakefulness than during rapid eye movement sleep. *J Neurosci* 2011;**31**(7):2649–56.
124. Xi MC, Morales FR, Chase MH. Evidence that wakefulness and REM sleep are controlled by a GABAergic pontine mechanism. *J Neurophysiol* 1999;**82**(4):2015–19.
125. Vanini G, Watson CJ, Lydic R, Baghdoyan HA. Gamma-aminobutyric acid-mediated neurotransmission in the pontine reticular formation modulates hypnosis, immobility, and breathing during isoflurane anesthesia. *Anesthesiology* 2008;**109**(6):978–88.
126. Anaclet C, et al. Genetic activation, inactivation, and deletion reveal a limited and nuanced role for somatostatin-containing basal forebrain neurons in behavioral state control. *J Neurosci* 2018;**38**(22):5168–81.
127. Kim T, et al. Cortically projecting basal forebrain parvalbumin neurons regulate cortical gamma band oscillations. *Proc Natl Acad Sci USA* 2015;**112**(11):3535–40.
128. Herrera CG, et al. Hypothalamic feedforward inhibition of thalamocortical network controls arousal and consciousness. *Nat Neurosci* 2016;**19**(2):290–8.
129. Venner A, Anaclet C, Broadhurst RY, Saper CB, Fuller PM. A novel population of wake-promoting GABAergic neurons in the ventral lateral hypothalamus. *Curr Biol* 2016;**26**(16):2137–43.
130. Lee RS, Steffensen SC, Henriksen SJ. Discharge profiles of ventral tegmental area GABA neurons during movement, anesthesia, and the sleep-wake cycle. *J Neurosci* 2001;**21**(5):1757–66.
131. Von Economo C. Sleep as a problem of localization. *J Nerv Ment Dis* 1930;71249–59.
132. Kumar VM. Why the medial preoptic area is important for sleep regulation. *Indian J Physiol Pharmacol* 2004;**48**(2):137–49.

133. Torterolo P, Benedetto L, Lagos P, Sampogna S, Chase MH. State-dependent pattern of Fos protein expression in regionally-specific sites within the preoptic area of the cat. *Brain Res* 2009;126744–56.
134. Szymusiak R, Gvilia I, McGinty D. Hypothalamic control of sleep. *Sleep Med* 2007;**8**(4):291–301.
135. Benedetto L, Chase M, Torterolo P. GABAergic processes within the median preoptic nucleus promote NREM sleep. *Behav Brain Res* 2012;23260–5.
136. Sherin JE, Elmquist JK, Torrealba F, Saper CB. Innervation of histaminergic tuberomammillary neurons by GABAergic and galaninergic neurons in the ventrolateral preoptic nucleus of the rat. *J Neurosci* 1998;**18**(12):4705–21.
137. McGinty D, Szymusiak R. Sleep-promoting mechanisms in mammals. In: Kryger MH, et al., editors. *Principles and practices of sleep medicine*. Philadelphia, PA: Elsevier-Saunders; 2005. p. 169–84.
138. Chung S, et al. Identification of preoptic sleep neurons using retrograde labelling and gene profiling. *Nature* 2017;**545**(7655):477–81.
139. Saito YC, et al. GABAergic neurons in the preoptic area send direct inhibitory projections to orexin neurons. *Front Neural Circuits* 2013;7192.
140. Gallopin T, et al. Identification of sleep-promoting neurons in vitro. *Nature* 2000;**404**(6781):992–5.
141. Batini C, Moruzzi G, Palestini M, Rossi GF, Zanchetti A. Persistent patterns of wakefulness in the pretrigeminal midpontine preparation. *Science* 1958;**128**(3314):30–2.
142. Anaclet C, et al. Identification and characterization of a sleep-active cell group in the rostral medullary brainstem. *J Neurosci* 2012;**32**(50):17970–6.
143. Anaclet C, et al. The GABAergic parafacial zone is a medullary slow wave sleep-promoting center. *Nat Neurosci* 2014;**17**(9):1217–24.
144. Sakai K. Are there sleep-promoting neurons in the mouse parafacial zone? *Neuroscience* 2017;**367**:98–109.
145. Gelegen C, et al. Excitatory pathways from the lateral habenula enable propofol-induced sedation. *Curr Biol* 2018;**28**(4). 580-587e5.
146. Torterolo P, Vanini G. Nuevos conceptos sobre la generación y el mantenimiento de la vigilia. *Rev Neurol* 2010;**50**(12):747–58.
147. Monti J, Torterolo P, Pandi Perumal SR. The effects of benzodiazepine and nonbenzodiazepine agents, ramelteon, low-dose doxepin, suvorexant, and selective serotonin 5-HT2A receptor antagonists and inverse agonists on sleep and wakefulness. *Clin Med Insights* 2016;829–36.
148. Vanini G, Torterolo P, Baghdoyan H, Lydic R. Effects of general anesthetics on sleep-wake centers. In: Mashour GA, Lydic R, editors. *The neuroscientific foundations of anesthesiology*. Oxford: Oxford University Press; 2011.
149. Siegel JM. REM sleep. In: Kryger MH, et al., editors. *Principles and practices of sleep medicine*. Philadelphia, PA: Elsevier-Saunders; 2011. p. 92–111.
150. McCarley RW. Neurobiology of REM and NREM sleep. *Sleep Med* 2007;**8**(4):302–30.
151. Luppi PH, et al. *Paradoxical (REM) sleep genesis: the switch from an aminergic-cholinergic to a GABAergic-glutamatergic hypothesis. J Physiol (Paris)*. 2007. p. 100271–83.
152. McCarley RW, Hobson JA. Neuronal excitability modulation over the sleep cycle: a structural and mathematical model. *Science* 1975;**189**(4196):58–60.
153. Crochet S, Sakai K. Effects of microdialysis application of monoamines on the EEG and behavioural states in the cat mesopontine tegmentum. *Eur J Neurosci* 1999;**11**(10):3738–52.
154. Luebke JI, et al. Serotonin hyperpolarizes cholinergic low-threshold burst neurons in the rat laterodorsal tegmental nucleus in vitro. *Proc Natl Acad Sci USA* 1992;**89**(2):743–7.

155. Williams JA, Reiner PB. Noradrenaline hyperpolarizes identified rat mesopontine cholinergic neurons in vitro. *J Neurosci* 1993;**13**(9):3878–83.
156. Van Dort CJ, et al. Optogenetic activation of cholinergic neurons in the PPT or LDT induces REM sleep. *Proc Natl Acad Sci USA* 2015;**112**(2):584–9.
157. Kubin L. Carbachol models of REM sleep: recent developments and new directions. *Arch Ital Biol* 2001;**139**(1-2):147–68.
158. Chase M, Morales FR. Control of motoneurons during sleep. In: Kryger MH, et al., editors. *Principles and practices of sleep medicine.* Philadelphia, PA: Elsevier-Saunders; 2005. p. 154–68.
159. Weng FJ, et al. Carbachol excites sublaterodorsal nucleus neurons projecting to the spinal cord. *J Physiol* 2014;**592**(7):1601–17.
160. Clement O, Sapin E, Berod A, Fort P, Luppi PH. Evidence that neurons of the sublaterodorsal tegmental nucleus triggering paradoxical (REM) sleep are glutamatergic. *Sleep* 2011;**34**(4):419–23.
161. Sapin E, et al. Localization of the brainstem GABAergic neurons controlling paradoxical (REM) sleep. *PLoS One* 2009;**4**(1):e4272.
162. Torterolo P, Yamuy J, Sampogna S, Morales FR, Chase MH. GABAergic neurons of the laterodorsal and pedunculopontine tegmental nuclei of the cat express c-fos during carbachol-induced active sleep. *Brain Res* 2001;**892**(2):309–19.
163. Torterolo P, Yamuy J, Sampogna S, Morales FR, Chase MH. GABAergic neurons of the cat dorsal raphe nucleus express c-fos during carbachol-induced active sleep. *Brain Res* 2000;**884**(1–2):68–76.
164. Torterolo P, Sampogna S, Morales FR, Chase MH. Gudden's dorsal tegmental nucleus is activated in carbachol-induced active (REM) sleep and active wakefulness. *Brain Res* 2002;**944**(1–2):184–9.
165. Torterolo P, Morales FR, Chase MH. GABAergic mechanisms in the pedunculopontine tegmental nucleus of the cat promote active (REM) sleep. *Brain Res* 2002;**944**(1-2):1–9.
166. Vanini G, Torterolo P, McGregor R, Chase MH, Morales FR. GABAergic processes in the mesencephalic tegmentum modulate the occurrence of active (rapid eye movement) sleep in guinea pigs. *Neuroscience* 2007;1451157–67.
167. Torterolo P, Vanini G. Involvement of GABAergic mechanisms in the laterodorsal and pedunculopontine tegmental nuclei (LDT-PPT) in the promotion of REM sleep. In: Monti J, et al., editors. *GABA and sleep: molecular, functional and clinical aspects.* Springer: Basel; 2010. p. 213–31.
168. Weber F, et al. Regulation of REM and Non-REM sleep by periaqueductal GABAergic neurons. *Nat Commun* 2018;**9**(1):354.
169. Grace KP, Horner RL. Evaluating the evidence surrounding pontine cholinergic involvement in REM sleep generation. *Front Neurol* 2015;6190.
170. Gillin JC, Sitaram N. Rapid eye movement (REM) sleep: cholinergic mechanisms. *Psychol Med* 1984;**14**(3):501–6.
171. Chase MH. Motor control during sleep and wakefulness: clarifying controversies and resolving paradoxes. *Sleep Med Rev* 2013;**17**(4):299–312.
172. Brooks PL, Peever JH. Identification of the transmitter and receptor mechanisms responsible for REM sleep paralysis. *J Neurosci* 2012;**32**(29):9785–95.
173. Weber F, et al. Control of REM sleep by ventral medulla GABAergic neurons. *Nature* 2015;**526**(7573):435–8.
174. Goutagny R, et al. Role of the dorsal paragigantocellular reticular nucleus in paradoxical (rapid eye movement) sleep generation: a combined electrophysiological and anatomical study in the rat. *Neuroscience* 2008;**152**(3):849–57.
175. Solms M. Dreaming and REM sleep are controlled by different brain mechanisms. *Behav Brain Sci* 2000;**23**(6):843–50 discussion 904–1121.

176. Lu J, et al. Selective activation of the extended ventrolateral preoptic nucleus during rapid eye movement sleep. *J Neurosci* 2002;**22**(11):4568–76.
177. Lee MG, Hassani OK, Jones BE. Discharge of identified orexin/hypocretin neurons across the sleep-waking cycle. *J Neurosci* 2005;**25**(28):6716–20.
178. Mileykovskiy BY, Kiyashchenko LI, Siegel JM. Behavioral correlates of activity in identified hypocretin/orexin neurons. *Neuron* 2005;**46**(5):787–98.
179. Torterolo P, Chase MH. The hypocretins (orexins) mediate the "phasic" components of REM sleep: a new hypothesis. *Sleep Sci* 2014;719–29.
180. Jakovljevic M, Sagud M, Mihaljevic-Peles A. Olanzapine in the treatment-resistant, combat-related PTSD – a series of case reports. *Acta Psychiatr Scand* 2003;**107**(5):394–6 discussion 396.
181. Thompson DF, Pierce DR. Drug-induced nightmares. *Ann Pharmacother* 1999;**33**(1):93–8.
182. Torterolo P, Sampogna S, Morales FR, Chase MH. MCH-containing neurons in the hypothalamus of the cat: searching for a role in the control of sleep and wakefulness. *Brain Res* 2006;1119101–14.
183. Torterolo P, Lagos P, Monti JM. Melanin-concentrating hormone (MCH): a new sleep factor? *Front Neurol* 2011;**2**(14):1–12.
184. Monti JM, Torterolo P, Lagos P. Melanin-concentrating hormone control of sleep-wake behavior. *Sleep Med Rev* 2013;17293–8.
185. Bittencourt JC, Diniz GB. Neuroanatomical structure of the MCH system. In: Pandi Perumal SR, et al., editors. *Melanin-concentrating hormone and sleep*. Switzerland: Springer; 2018.
186. Costa A, Castro-Zaballa S, Lagos P, Chase MH, Torterolo P. Distribution of MCH-containing fibers in the feline brainstem: relevance for REM sleep regulation. *Peptides* 2018.
187. Tan CP, et al. Melanin-concentrating hormone receptor subtypes 1 and 2: species-specific gene expression. *Genomics* 2002;**79**(6):785–92.
188. Hassani OK, Lee MG, Jones BE. Melanin-concentrating hormone neurons discharge in a reciprocal manner to orexin neurons across the sleep-wake cycle. *Proc Natl Acad Sci USA* 2009;**106**(7):2418–22.
189. Blanco-Centurion C, Liu M, Shiromani P. Optogenetic control of the melanin-concentrating hormone expressing neurons. In: Pandi Perumal SR, et al., editors. *Melanin-concentrating hormone and sleep*. Switzerland: Springer; 2018. p. 75–108.
190. Vetrivelan R, et al. Melanin-concentrating hormone neurons specifically promote rapid eye movement sleep in mice. *Neuroscience* 2016;336102–13.
191. Lagos P, Torterolo P, Jantos H, Chase MH, Monti JM. Effects on sleep of melanin-concentrating hormone microinjections into the dorsal raphe nucleus. *Brain Res* 2009;1265103–10.
192. Torterolo P, Sampogna S, Chase MH. MCHergic projections to the nucleus pontis oralis participate in the control of active (REM) sleep. *Brain Res* 2009;**1268**:76–87.
193. Verret L, et al. A role of melanin-concentrating hormone producing neurons in the central regulation of paradoxical sleep. *BMC Neurosci* 2003;**4**(1):19.
194. Monti JM, Lagos P, Jantos H, Torterolo P. Increased REM sleep after intra-locus coeruleus nucleus microinjection of melanin-concentrating hormone (MCH) in the rat. *Prog Neuropsychopharmacol Biol Psychiatry* 2015;56185–8.
195. Benedetto L, et al. Microinjection of melanin concentrating hormone into the lateral preoptic area promotes non-REM sleep in the rat. *Peptides* 2013;**39**:11–15.
196. Lagos P, Monti JM, Jantos H, Torterolo P. Microinjection of the melanin-concentrating hormone into the lateral basal forebrain increases REM sleep and reduces wakefulness in the rat. *Life Sci* 2012;**90**(23–24):895–9.

197. Urbanavicius J, Lagos P, Torterolo P, Abin-Carriquiry JA, Scorza C. Melanin-concentrating hormone projections to the dorsal raphe nucleus: an immunofluorescence and in vivo microdialysis study. *J Chem Neuroanat* 2016;**72**:16–24.
198. Devera A, et al. Melanin-concentrating hormone (MCH) modulates the activity of dorsal raphe neurons. *Brain Res* 2015;1598114–28.
199. Lagos P, Urbanavicius J, Scorza C, Miraballes R, Torterolo P. Depressive-like profile induced by MCH microinjections into the dorsal raphe nucleus evaluated in the forced swim test. *Behav Brain Res* 2011;218259–66.
200. Urbanavicius J, Lagos P, Torterolo P, Scorza C. Pro-depressive effect induced by microinjections of MCH into the dorsal raphe: time-course, dose-dependence, effects on anxiety-related behaviors and reversion by nortriptyline. *Behav Pharmacol* 2014;25316–24.
201. Lopez Hill X, Pascovich C, Urbanavicius J, Torterolo P, Scorza C. The median raphe nucleus participates in the depressive-like behavior induced by MCH: differences with the dorsal raphe nucleus. *Peptides* 2013;**50**:96–9.
202. Torterolo P, et al. Melanin-concentrating hormone (MCH): role in REM sleep and depression. *Front Neurosci* 2015;**9**:475.
203. Urbanavicius J, Lagos P, Lopez-Hill X, Torterolo P, Scorza C. MCH and depression. In: Pandi Perumal SR, et al., editors. *Melanin-concentrating hormone and sleep*. Switzerland: Springer; 2018. p. 195–206.
204. Varin C, Luppi PH, Fort P. Melanin-Concentrating Hormone-expressing neurons adjust slow-wave sleep dynamics to catalyze paradoxical (REM) sleep. *Sleep* 2018;**41**.
205. Chaki S. MCH receptor 1 antagonists: antidepressant/anxiolytic potential in animal models. In: Pandi Perumal SR, et al., editors. *Melanin-concentrating hormone and sleep*. Switzerland: Springer; 2018. p. 207–25.
206. Pandi-Perumal SR, et al. Physiological effects of melatonin: role of melatonin receptors and signal transduction pathways. *Prog Neurobiol* 2008;**85**(3):335–53.
207. Scheer FA, Czeisler CA. Melatonin, sleep, and circadian rhythms. *Sleep Med Rev* 2005;**9**(1):5–9.
208. Sharma R, Sahota P, Thakkar MM. Melatonin promotes sleep in mice by inhibiting orexin neurons in the perifornical lateral hypothalamus. *J Pineal Res* 2018; e12498.
209. Schmitt LI, Sims RE, Dale N, Haydon PG. Wakefulness affects synaptic and network activity by increasing extracellular astrocyte-derived adenosine. *J Neurosci* 2012;**32**(13):4417–25.
210. Porkka-Heiskanen T, Alanko L, Kalinchuk A, Stenberg D. Adenosine and sleep. *Sleep Med Rev* 2002;**6**(4):321–32.
211. Strecker RE, et al. Adenosinergic modulation of basal forebrain and preoptic/anterior hypothalamic neuronal activity in the control of behavioral state. *Behav Brain Res* 2000;**115**(2):183–204.
212. Gallopin T, et al. The endogenous somnogen adenosine excites a subset of sleep-promoting neurons via A2A receptors in the ventrolateral preoptic nucleus. *Neuroscience* 2005;**134**(4):1377–90.
213. Lazarus M, Huang ZL, Lu J, Urade Y, Chen JF. How do the basal ganglia regulate sleep-wake behavior? *Trends Neurosci* 2012;**35**(12):723–32.
214. Lazarus M, et al. Arousal effect of caffeine depends on adenosine A2A receptors in the shell of the nucleus accumbens. *J Neurosci* 2011;**31**(27):10067–75.
215. Oishi Y, et al. Slow-wave sleep is controlled by a subset of nucleus accumbens core neurons in mice. *Nat Commun* 2017;**8**(1):734.
216. Nishino S, Mignot E. Wake-promoting medications: basic mechanisms and pharmacology. In: Kryger MH, et al., editors. *Principles and practices of sleep medicine*. Philadelphia, PA: Elsevier-Saunders; 2005. p. 468–83.

217. Roehrs T, Roth T. Caffeine: sleep and daytime sleepiness. *Sleep Med Rev* 2008;**12**(2):153–62.
218. Falconi A, et al. Waking-promoting action of the Yerba Mate (*Ilex paraguariensis*). *Sleep Science* 2013;**6**:9–15.
219. Torterolo P, et al. Yerba Mate: efectos sobre la vigilia y el sueño. *Ann Facul Med* 2014;**1**(1):24–7.
220. Imeri L, Opp MR. How (and why) the immune system makes us sleep. *Nat Rev Neurosci* 2009;**10**(3):199–210.
221. Krueger JM, et al. Involvement of cytokines in slow wave sleep. *Prog Brain Res* 2011;19339–47.
222. Krueger JM, Opp MR. Sleep and microbes. *Int Rev Neurobiol* 2016;**131**:207–25.
223. Murillo-Rodriguez E. The modulatory role of endocannabinoids in sleep. *Rev Neurol* 2008;**46**(3):160–6.
224. Pava MJ, Makriyannis A, Lovinger DM. Endocannabinoid signaling regulates sleep stability. *PLoS One* 2016;**11**(3):e0152473.
225. Russo E. *Cannabis in India: ancient lore and modern medicine, Cannabinoids as therapeutics. Milestones in drug therapy*. Basel, Boston: Birkhäuser; 2005. p. 1–22.
226. Vincent SR, Satoh K, Armstrong DM, Fibiger HC. NADPH-diaphorase: a selective histochemical marker for the cholinergic neurons of the pontine reticular formation. *Neurosci Lett* 1983;**43**(1):31–6.
227. Marino J, Cudeiro J. Nitric oxide-mediated cortical activation: a diffuse wake-up system. *J Neurosci* 2003;**23**(10):4299–307.
228. Monti JM, Jantos H. Effects of L-arginine and SIN-1 on sleep and waking in the rat during both phases of the light-dark cycle. *Life Sci* 2004;**75**(17):2027–34.
229. Monti JM, Jantos H. Microinjection of the nitric oxide synthase inhibitor L-NAME into the lateral basal forebrain alters the sleep/wake cycle of the rat. *Prog Neuropsychopharmacol Biol Psychiatry* 2004;**28**(2):239–47.
230. Kapas L, Krueger JM. Nitric oxide donors SIN-1 and SNAP promote nonrapid-eye-movement sleep in rats. *Brain Res Bull* 1996;**41**(5):293–8.
231. Datta S, Patterson EH, Siwek DF. Endogenous and exogenous nitric oxide in the pedunculopontine tegmentum induces sleep. *Synapse* 1997;**27**(1):69–78.
232. Leonard TO, Lydic R. Pontine nitric oxide modulates acetylcholine release, rapid eye movement sleep generation, and respiratory rate. *J Neurosci* 1997;**17**(2):774–85.
233. Kalinchuk AV, Stenberg D, Rosenberg PA, Porkka-Heiskanen T. Inducible and neuronal nitric oxide synthases (NOS) have complementary roles in recovery sleep induction. *Eur J Neurosci* 2006;**24**(5):1443–56.
234. Gerashchenko D, et al. Identification of a population of sleep-active cerebral cortex neurons. *Proc Natl Acad Sci USA* 2008;**105**(29):10227–32.
235. Morairty SR, et al. A role for cortical nNOS/NK1 neurons in coupling homeostatic sleep drive to EEG slow wave activity. *Proc Natl Acad Sci USA* 2013;**110**(50):20272–7.
236. Dittrich L, Morairty SR, Warrier DR, Kilduff TS. Homeostatic sleep pressure is the primary factor for activation of cortical nNOS/NK1 neurons. *Neuropsychopharmacology* 2015;**40**(3):632–9.
237. Cespuglio R, Debilly G, Burlet S. Cortical and pontine variations occurring in the voltammetric no signal throughout the sleep-wake cycle in the rat. *Arch Ital Biol* 2004;**142**(4):551–6.
238. Singh C, Rihel J, Prober DA. Neuropeptide Y regulates sleep by modulating noradrenergic signaling. *Curr Biol* 2017;**27**(24). 3796–3811e5.
239. Lee DA, et al. Genetic and neuronal regulation of sleep by neuropeptide VF. *Elife* 2017;6.
240. Richter C, Woods IG, Schier AF. Neuropeptidergic control of sleep and wakefulness. *Annu Rev Neurosci* 2014;**37**:503–31.

241. Oliveira FT, et al. Altered sleep patterns in patients with non-functional GHRH receptor. *Eur J Endocrinol* 2017;**177**(1):51–7.
242. Deurveilher S, Rusak B, Semba K. Female reproductive hormones alter sleep architecture in ovariectomized rats. *Sleep* 2011;**34**(4):519–30.
243. Jego S, et al. Tuberal hypothalamic neurons secreting the satiety molecule Nesfatin-1 are critically involved in paradoxical (REM) sleep homeostasis. *PLoS One* 2012;**7**(12): e52525.
244. Larson-Prior LJ, Ju YE, Galvin JE. Cortical-subcortical interactions in hypersomnia disorders: mechanisms underlying cognitive and behavioral aspects of the sleep-wake cycle. *Front Neurol* 2014;**5**:165.
245. Wisor J. Modafinil as a catecholaminergic agent: empirical evidence and unanswered questions. *Front Neurol* 2013;**4**:139.
246. Nishino S, Mignot E. Wake-promoting medications: basic mechanisms and pharmacology. In: Krieger MH, et al., editors. *Principles and practices of sleep medicine.* Philadelphia, PA: Saunders; 2011. p. 510–26.
247. Syed YY. Pitolisant: first global approval. *Drugs* 2016;**76**(13):1313–18.
248. Calik MW. Update on the treatment of narcolepsy: clinical efficacy of pitolisant. *Nat Sci Sleep* 2017;**9**:127–33.
249. Pandi-Perumal SR, BaHammam AS, Shapiro CM. Parasomnias. In: Stolerman IP, Price LH, editors. *Encyclopedia of psychopharmacology*. Berlin Heidelberg: Springer-Verlag; 2014.
250. Mahowald MW, Schneck C. REM sleep parasomnias. In: Kryger MH, et al., editors. *Principles and practices of sleep medicine*. Philadelphia, PA: Elsevier-Saunders; 2011. p. 1083–97.
251. Reid KJ, Zee PC. Circadian disorders of the sleep-wake cycle. In: Kryger MH, et al., editors. *Principles and practices of sleep medicine*. Philadelphia, PA: Elsevier-Saunders; 2011. p. 470–82.
252. Reid KJ, Abbott SM. Jet lag and shift work disorder. *Sleep Med Clin* 2015;**10** (4):523–35.
253. Mayers AG, Baldwin DS. Antidepressants and their effect on sleep. *Hum Psychopharmacol* 2005;**20**(8):533–59.
254. Wichniak A, Wierzbicka A, Jernajczyk W. Sleep and antidepressant treatment. *Curr Pharm Des* 2012;**18**(36):5802–17.
255. Everitt H, et al. Antidepressants for insomnia in adults. *Cochrane Database Syst Rev* 2018;**5** CD010753.
256. Gandhi KR, Saadabadi A. *Levodopa (L-Dopa)*. Treasure Island, FL: StatPearls; 2018.
257. Loddo G, et al. The treatment of sleep disorders in Parkinson's disease: from research to clinical practice. *Front Neurol* 2017;**8**:42.
258. Atkin T, Comai S, Gobbi G. Drugs for insomnia beyond benzodiazepines: pharmacology, clinical applications, and discovery. *Pharmacol Rev* 2018;**70**(2):197–245.

CHAPTER

5

The Role of Neuroglobin in Brain Function and Sleep—Wake Cycle

Oscar Ramírez-Plascencia[1,2], *Mario Eduardo Acosta-Hernández*[1] *and Fabio García-García*[1]

[1]Biomedicine Department, Health Sciences Institute, Veracruzana University, Veracruz, Mexico [2]Medicine School, Autonomous University of San Luis Potosí, San Luis Potosi, Mexico

OUTLINE

Molecular Sleep—Wake Cycle
DOI: https://doi.org/10.1016/B978-0-12-816430-3.00005-1

5.1 INTRODUCTION

Globins are a family of small globular metalloproteins of about 150 amino acids in length and characterized by 8 α-helical segments (A-H) that exert a characteristic 3-over-3 folding, which is known as the globular structure.[1,2] They contain a heme prosthetic group (Fe-protoporphyrin XI) that is able to form reversible binding with gaseous ligands as oxygen (O_2), carbon dioxide (CO_2), or nitric oxide (NO).[2] Phylogenetically, globins are ancient molecules and have been reported in bacteria, fungi, plants, vertebrates, and invertebrates, all related to respiratory function, supplying O_2 to the tissues and storing intracellular O_2.[2,3]

The globin family includes eight different types of globins, but the best described and characterized in vertebrates are hemoglobin (Hb), myoglobin (Mb), cytoglobin (Cygb), and neuroglobin (Ngb). Hb is located in erythrocytes, which transport O_2 from the lungs to tissues through the circulatory system and CO_2 in the opposite direction. Mb is mainly located in skeletal and cardiac muscle where it stores and facilitates intracellular O_2 diffusion to the mitochondria, and exerts an important role in NO scavengers, biotransforming NO radicals in nitrate. Cygb is ubiquitously expressed in all tissues and has been related with collagen synthesis and protection against hypoxia. Finally, Ngb is located in the central nervous system (CNS), mainly in neurons, and is suggested to take part in different neuroprotective mechanisms.[2–4] In this chapter we discuss Ngb's roles in brain function and, particularly, its role in sleep/wake cycle regulation.

5.2 NEUROGLOBIN

Ngb is a monomeric globular hemoprotein of about 150 amino acids with a relative molecular mass of 17 kDa. First described in the mammal brain in 2000, it has been reported widely in vertebrates from human, mouse, rat, to blowfish and zebrafish, which suggests that Ngb protein is highly conserved through the evolution.[4–6] Ngb is codified by a single copy gene in all species, and the coding region in mammalian Ngb gene carries three introns situated at B12.2, E11.0, and G7.0. In spite of the fact that introns B12.2 and G7.0 are present in other globin genes, the intron E11.0 is unique to Ngb.[7,8] All vertebrates contain these introns in a conserved position within the coding regions except zebrafish (*Danio rerio*), which contains a fourth intron.[7] The human Ngb gene is located on the long arm of chromosome 14 at position q24.3.[7] Ngb shares about 20%–25% sequence identity compared with the

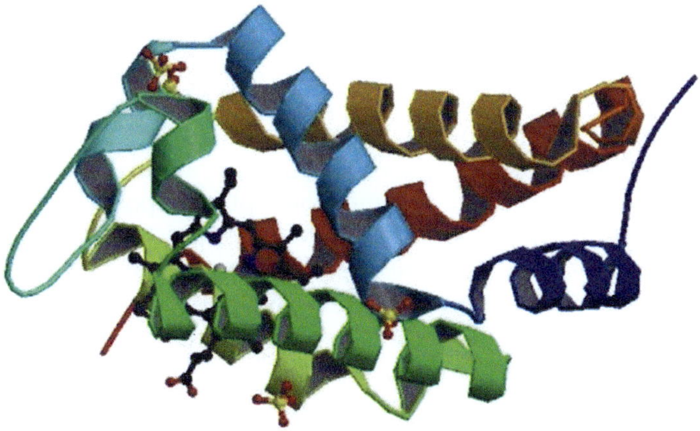

FIGURE 5.1 Three-dimensional structure of human brain neuroglobin. Structure reveals a distinct mode of controlling O_2 affinity, which stretches from the heme distal site toward the protein surface.[9]

homologous globins Hb and Mb[5,8]; despite the difference, Ngb adopts a 3D folding identical to the rest of the globins, which is the main characteristic of the family (Fig. 5.1).

Ngb shares almost the same structure with other globins. The main difference between them is in the following: Hb and Mb Fe^{2+} atom has six coordination bonds for six ligands, four of them are nitrogen atoms that are part of the porphyrin molecule, the fifth bond lies to the porphyrin with the histidine's nitrogen atom at position 93, while the sixth bond is with O_2. In contrast, Ngb forms six bonds, but all of them are destined for the O_2.[10] The occupancy of the six-coordinating bonds implies that exogenous ligands have to displace the distal histidine to bond to the prostatic group; for that reason, the mechanism of exogenous ligand binding to Ngb is more complex than the other globins.[10] Thus, Ngb may enhance O_2 to metabolic highly neurons and may also be involved in the detoxification of reactive oxygen species (ROS).[10,11]

5.3 NEUROGLOBIN DISTRIBUTION IN THE BRAIN

Ngb is expressed in the CNS and peripheral nervous system (PNS), mainly in neurons,[10] but also is observed in other endocrine tissues. Its concentration in CNS rates at the micromolar level, about 0.01% of the total protein content in the brain.[9] The distribution of Ngb in CNS varies depending on the species, but different evidence shows broad expression of mRNA and protein in different brain regions. In the first report by Burmester et al.[1] using Western blot analysis, the major expression of Ngb was detected in the subthalamic nucleus, with lower proportion in the brain cortex, thalamus and medulla oblongata, cerebellar cortex and substantia nigra, while in caudate nucleus and hippocampus 10% was observed compared with that expressed in subthalamic nucleus (Table 5.1). Since then, other reports have reported high expression of Ngb in the hypothalamus, particularly in the lateral hypothalamic area, paraventricular nucleus, arcuate nucleus, dorsomedial hypothalamic nucleus, the preoptic area (POA), and the

TABLE 5.1 Major Regions of the Brain Where Neuroglobin Is Expressed

Brain nucleus	Ngb expression	Related with sleep/wake cycle
Cortex	+	√
Thalamus	+	√
Medulla oblongata	+	
Substantia nigra	+	
Caudate nucleus	+	
Hippocampus	+	√
Cerebellum	+	
Locus coeruleus	+	√
Anterior basomedial amygdaloid nuclei	+	
Paraventricular nucleus	+++	
Preoptic area	+++	√
Lateral hypothalamic area	+++	√
Arcuate nucleus	+++	
Pedunculopontine tegmental nucleus	+++	√
Laterodorsal tegmental nucleus	+++	√
Dorsomedial hypothalamic nucleus	+++	
Posterodorsal medial amygdaloid nuclei	+++	
Subthalamic nucleus	+++	
Nucleus of the solitary tract	+++	

+ and +++ symbols represent lower and higher expression respectively. √ symbol indicates areas that express Ngb and are related with sleep/wake regulation.

mammillary region. Out of the hypothalamus, Ngb has been described in the laterodorsal tegmental nucleus (LDTg), pedunculopontine tegmental nucleus (PPTg), the anterior basomedial, posterodorsal medial amygdaloid nuclei, locus coeruleus (LC), and nucleus of the solitary tract (NTS) in brainstem, and low concentrations were observed in spinal cord and thymus. However, there is no expression in the hippocampus and weak expression in the cerebral cortex was observed.[5] Moreover, Hundahl and his collaborators did not detect Ngb in the cerebellum[12]; meanwhile Alekseeva reported in the human and rat cerebellum Ngb is expressed in Purkinje cells with moderate expression in the granular layer of cerebellar glomeruli (Table 5.1).[5,10]

Ngb has been reported in some tissues outside of the CNS. A large amount of Ngb was detected in the carotid bodies[12] and visual systems; for example, an immunohistochemical study carried out in mouse, dog, and human revealed Ngb expression in the ganglion cell layer, inner nuclear layer, and outer plexiform layers and pigment epithelium of the retina,[5,13–15] besides in the corneal epithelium, endothelium, and iris and in the corner of the eyeball's anterior chamber.[5,16] In addition, Ngb was also detected in the endocrine system in the gastrointestinal tract such as in the pancreas, stomach, and intestines; moreover, it has been detected in testicles, lymph nodes, and ovaries.[5] In mouse and rat it was detected in the esophagus, duodenum, large intestine, enteroendocrine cells, anterior pituitary, adrenals, testicular spermatogonia and spermatocytes, and later was identified in cardiomyocytes.[1,5] Remarkably, no expression has been found in heart, aorta, spleen, skeletal muscle, urinary bladder, uterus, prostate, thyroid, mammary, and salivary glands in humans.[1,5]

5.4 THE ROLE OF NEUROGLOBIN IN BRAIN FUNCTION

The first description by Burmester and collaborators[1] initiated extensive research into the potential function of Ngb. As expected, Ngb was initially related to the maintenance and regulation of cellular oxygen supply in neurons, similar to the physiological function of other globins, which was logical given the elevated active metabolism that neurons have. However, the evidence does not completely support this hypothesis, and the description of low concentration of the protein in physiological condition (below 0.01% of total protein content in brain)[1] or the little or no effect when the protein is absent or reduced,[17–19] led to a search for its main function to other directions. Although the debate about Ngb's main function is still on the table, it is broadly accepted that it serves as a neuroprotective protein in different cellular stresses or damage. In that sense, Ngb is a highly stable protein and maintains its structural properties in different conditions,[20] which is an important characteristic in the case of brain damage, and to this objective, its levels are upregulated in many brain insults.

The first evidence of Ngb as neuroprotectant protein derives from works where oxygen saturation was compromised. Just 1 year after its description, Sun and collaborators[21] showed an increase in Ngb protein and mRNA after hypoxic-ischemic injury in mouse cerebral neuron cultures. After that, other reports confirmed Ngb increases when oxygen saturation is reduced in vitro and in vivo[22–25]; even in humans after ischemic stroke the increment was reported,[26] and along with Ngb increments, protective effects were described. Nevertheless, some reports have also failed to observe these changes,[27–29] which poses the question of whether the response is dependent on specific conditions. The hypoxia triggers different signals that terminate in protein expression to avoid cell damage or apoptosis, and the main transcriptional regulator under hypoxic condition is hypoxia-inducible factor 1 (HIF1). As Ngb levels positively correlate with HIF1,[30,31] the first was considered as part of the downstream regulated proteins of the second. In support of this idea, the HIF1 pharmacological induction with tibolone exerts a favorable effect in the CNS and increases Ngb levels; meanwhile Ngb silencing avoids this and reduces the cell survival under oxygen–glucose deprivation (OGD).[32,33] However, no HIF1 motive was found in Ngb promoter, but other different pathways implied in hypoxia-mediated gene regulation sites were described, such as hypoxia-inducible protein binding, Sp1, AP1, NF-1, and NFÎB motives.[34] Thus, alternative pathways of how HIF1 controls Ngb were proposed, where the principal is the induction of Ngb expression through Sp1,[30,35,36] although other works have pointed out that vascular endothelial growth factor and erythropoietin, which also are HIF-1 targets in hypoxia conditions,[37] may mediate Ngb expression.[38,39] Therefore, Ngb increments seem to be necessary to reduce damage in cell stress by the reduction of oxygen saturation, and it is downstream regulated by HIF-1 through Sp1 or alternative pathways.

The high affinity for ROS and NO suggested that Ngb was related with their homeostasis regulation, mainly when they are produced during damage. For example, after ischemia-reperfusion, the oxygen saturation decays and an overproduction of ROS and NO is triggered. In hypoxia or ischemia-reperfusion NO exerts beneficial effects acting as a vasodilator, but an excess terminates in neurotoxicity because it interacts with superoxides and other ROS, so NO levels need to be limited. In that context, Ngb regulates NO levels to constrain oxidative molecules restricting mitochondrial respiration in hypoxia conditions. In that sense, the association of Ngb with mitochondria is vital since it is the major generator of ROS; while Ngb expression was mainly reported in cytosol, a significant fraction (close to 8%) is expressed in mitochondria,[40–43] but this may reach up to 70% in retinal neurons of rats,[41] and further, translocation of cytosolic Ngb to mitochondria occurs after a brain insult as OGD.[40,44] The location around and within mitochondria allows Ngb to be one important scavenger of NO and ROS regulation, mainly containing damage after a brain insult.[45–47] The relation with mitochondria is also functional since Ngb overexpression reduces the impact of hypoxia in ATP production and mitochondria membrane potential in primary culture neurons,[48] meanwhile a knockdown in retinal Ngb reduces the activity of complex I and III in the mitochondrial respiratory chain.[41] Hence, the interaction of Ngb with the mitochondria is not only significant in neuroprotection but also in physiological mitochondrial functions.

The neuroprotectant effects of Ngb also include an important role preventing apoptosis; for this propose, the interaction with mitochondrial proteins is essential. For example, Ngb interacts directly with cytochrome C (Cyc)[44,49] and electron transporter protein between complex III and IV in the mitochondrial respiratory chain, but when cellular damage accumulates it may be released to the cytosol where it interacts with apoptotic protease activating factor 1 (APAF-1) to activate caspase 9 for the induction of the apoptosome. In this sense, one in silico study pointed out that Ngb reorganizes Lys72 and Lys25 residues in Cyc required for its binding to APAF-1.[50] The antiapoptotic effects of Ngb through Cyc regulation are supported by evidence as the levels of Ngb are associated to Cyc increases as a response of OGD,[44] or when Ngb is overexpressed, reducing Cyc release from the mitochondria.[51] The reduction of mitochondria Cyc release by overexpression of Ngb improves functional recovery after spinal cord injury, and promotes less neuronal loss and fewer apoptotic cells.[52] Besides, Ngb interacts with voltage-dependent anion channel (VDAC),[40,51] a protein involved in ATP and Ca^{2+} transport in the mitochondrial outer membrane, but after a cellular insult, participates in the mitochondrial permeability transition (MPT), allowing the release of small molecules as Cyc through a MPT pore (mPTP).[53] This interaction is concordant with antiapoptotic function since Ngb overexpression reduces the mPTP opening; meanwhile Ngb knockdown potentiates its opening in neurons and isolated mitochondria.[51] Interestingly, when VDAC is blocked, the Ngb translocation to mitochondria is inhibited.[40] Moreover, Ngb reduces the formation of lipid raft mitochondrial-actin cytoskeletal lattice death signal inhibiting Pak1 kinase activity and Rac1-GDP-dissociation,[54] in this case acting as a signaling molecule. Positive effects of Ngb reducing apoptosis are extended to the increase of antiapoptotic protein Bcl-2 and the decrease of proapoptotic protein Bax.[49,52,55] Thus, it is accepted that Ngb prevent apoptosis inhibiting Cyc release, interacting with other proapoptotic proteins as VDAC, participating in signaling pathways necessary in the formation of the apoptosome, and further, it is thought that the Ngb/Cyc ratio may set the threshold for apoptosis initiation.[15,56,57]

Ngb function extends to participate and regulate different intracellular signaling pathways. The clearest example in this case is that it prevents GTP/GDP exchange and the activation of the Gα-subunit in G-protein-coupled receptors, while activating phosphatidyl inositol 3-kinase (PI3K) through the Gβγ pathway.[58,59] In that line, Ngb increases PI3K phosphorylation of Akt, a situation that serves to preserve neuron cell viability under hydrogen peroxide insult and activate the mitochondrial K_{ATP} channel.[60] This pathway also is involved in apoptosis since it is observed by the activation of PI3K/Akt pathway in Ngb overexpression in mice that mitochondrial dysfunction and apoptosis are reduced.[61] Otherwise, Ngb also is involved in the regulation of AMP-activated protein kinase and the activity of acetyl-CoA carboxylase, promoting cellular anabolism through the regulation of ATP concentration.[62] Theoretically, Ngb may also be involved in cAMP response element binding (CREB) protein since present a site to respond in its promoter region.[63] CREB is associated with ROS detoxification and neuronal survival among other neuronal functions, and it is well described that the Akt pathway ends in CREB activation, so Ngb could trigger that signaling pathway and favor CREB's protective features, but experimental evidence is needed to support this hypothesis. More experiments are required for confirm the participation and mechanism of Ngb as a signaling molecule in brain injury, and also evaluate if it might participate in physiological conditions.

In neurodegenerative diseases as Alzheimer disease (AD), Ngb represents a cellular restraint in the first stages of the pathology development. The first observation indicated that Ngb is upregulated in AD patients,[64] probably induced by the mTOR-HIF1 pathway.[64,65] Increases in Ngb expression contain key events in AD development since it reduces the hyperphosphorylation of Tau,[66] attenuates cognitive impairment and decreases the generation of Aβ42 in rodent models[61]; all of these regulated through Akt pathway activation. Otherwise, in early AD stages, RE1-silencing transcription factor (REST) acts by repressing different proapoptotic genes.[67] Interestingly, Ngb promoter contains two REST sites[68] and REST transcription correlates with Ngb in a mouse model of AD,[69] which strongly suggests that this pathway represents a first line of protection against the initial events in early stages of AD. On the other hand, a conjugation of ibuprofen and (R)-α-lipoic acid is used in AD because it ameliorates the effects; for example, the conjugated administration in AD rats reduces apoptosome formation and cell death, and interestingly maintains Ngb levels similar to healthy controls as part of the antiapoptotic mechanisms.[70] The evidence of its participation in AD makes Ngb a possible pharmaceutical target to contain the pathological development in AD, and even though some complications in this aim are present as Ngb is membrane impermeable, alternative ways to induce Ngb increments may be a solution[33,70,71]; nevertheless more studies are required for this purpose.

Altogether, the evidence is strong in favor of Ngb's function as a neuroprotector. However, the Ngb response seems to be the first line of defense after cell damage, as observed in the increments at early stages of AD but not after, adding to reports that indicate an increase in the short term of the brain insult, but a decrease in the long term,[72,73] even in aging Ngb expression decay,[74] which is probably one of the reasons for aging as a factor risk in AD. What is clear is that independently of the brain insult or experimental protocol used, all data coincides that the regulation/inhibition of apoptosis by different pathways are at the center of Ngb's neuroprotective effects. Nevertheless, there are still some reports that fail to observe this effect, so more studies are needed to leave no doubt about Ngb's function and mechanism.

5.5 PARTICIPATION OF THE NEUROGLOBIN IN SLEEP–WAKE REGULATION CYCLE

Since Ngb is expressed in specific regions, part of its functions may be related to understanding the functions of these areas.[75,76] Particularly, Ngb is expressed in nuclei involved in sleep/wake cycle regulations, such as LC, PPTg, LDTg, and POA,[77] where the neurons modify their firing patterns across the sleep–wake cycle.[78] Recently, our work group demonstrated a reduction of the number of Ngb-positive cells in the nuclei PPTg, LDTg, and periventricular area in rats after sleep deprivation (SD) for 24 hours, which shows that the protein expression is sensible to alterations in sleep/wake homeostasis.[79] In that work, the plasma corticosterone levels in SD animals did not increase compared with non-SD rats, which suggests that the stress induced by SD was not a factor responsible for the change in Ngb expression, which implies that lack of sleep is the factor responsible for its reduction.[79] As Ngb exerts antioxidant effects,[80] we evaluated whether SD produces oxidative stress in brain tissue since insufficient sleep induced oxidative stress,[81,82] but the results showed that SD does not induce oxidative stress measured through the generation of malondialdehyde, a metabolite produced by membrane lipid peroxidation.[79] Altogether, the results support the hypothesis that the expression of Ngb depends on sleep homeostasis because the SD reduces its immunoreactivity even when no oxidative stress or physiological stress was observed. It is possible that Ngb expression could be regulated by the orexinergic pathway because orexins participate in the maintenance of wakefulness[83] and Ngb-positive neurons in LC, PPTg, and LDTg coexpress the orexin-A receptor.[84] In that sense, forced wakefulness induced by SD increases the OX-A levels into the brain and reduces the Ngb immunoreactivity, suggesting that the activation of the orexin receptors in these neurons may mediate the Ngb response to SD; however, that hypothesis should be investigated in detail.

The effects of sleep debt on Ngb expression have also been reported in obstructive sleep apnea (OSA), a condition that generates intermittent hypoxia (IH), sleep disruption, as well as increased apoptotic process due to oxidative stress. Interestingly, plasma levels of Ngb increase in proportion to the degree of severity in patients diagnosed with OSA.[85]

Furthermore, Ngb overexpression in IH animal models protects against lipid peroxidation,[86] and in a murine model of ischemia-reperfusion, Ngb positive cells notably increased 24-hour postreperfusion in the cerebral cortex and stay higher than controls for 48 hours after ischemia-reperfusion.[87] These findings provide evidence for the neuroprotective role of Ngb in cases like OSA. Sleep is also associated with neuroprotective functions since sleep debt is a risk factor to develop AD and other neurologic and psychiatry disorders.[64–66] As Ngb is also related with neurodegenerative diseases such as AD, and SD reduces the Ngb expression in sleep/wake related neurons, it is plausible that Ngb acts as the first line of defense in the development of neurodegenerative pathologies, and it may be one of the causes of why reducing sleep is a risk factor in these cases. Even though it sounds logical, it is necessary to investigate whether chronic sleep restriction may affect Ngb expression and if the reduction in Ngb affects the neurodegenerative disease's progress.

Ngb is highly expressed in neurons of the suprachiasmatic nucleus (SCN) in the hypothalamus, which serves as the master clock in the mammal brain. SCN neurons coexpress Ngb and PER-1, a clock gene involved in the circadian regulation and maintenance of the SCN activity.[88] SCN receives a monosynaptic input from the retina, where Ngb is highly expressed, and interestingly, Ngb mRNA expression in SCN neurons of mice increases during the light period, meanwhile during the dark period decreases, in a clear circadian pattern,[89] and moreover, neuroglobin deficient mice disrupt the SCN light responses and increase the expression of PER1 after a light stimuli during the dark period.[89] This also correlated with preliminary results of our laboratory showing an increase in the number of Ngb-positive cells during sleep (light phase) and decrease during wake phase (dark phase). These results suggest that the deficiency of Ngb does not affect the circadian clock function but its response to the light, additionally, suggests that Ngb is not only circadian regulated but also participates in light responses, and since sleep is regulated in circadian fashion directly from the SCN, it could be one of the main signals for Ngb synthesis and regulation.

5.6 CONCLUSIONS

Ngb is a monomeric hemoprotein expressed mainly in the brain of mammals and other vertebrates that exerts several functions in CNS, highlighting its participation in homeostatic mechanisms for O_2 and NO, intracellular signaling pathways, and regulating oxidative stress, mostly in cellular damage. Its neuroprotective effects are associated with apoptotic and energy-balance functions in a close interaction with mitochondria. However, new evidence suggests that Ngb is susceptible to response to the sleep–wake cycle and its alterations since the Ngb positive cells oscillate with the sleep/wake cycle and SD reduces the number of Ngb immunoreactivity in neurons associated with sleep regulation. Sleep regulation of Ngb may be related to orexin because it is well documented that forced wakefulness increases orexin levels and its receptors, associated also with a reduction of Ngb expression. Although the evidence clearly points to Ngb participation in the sleep/wake cycle, more studies are necessary to understand the mechanism of Ngb in sleep and other behaviors, and the consequences of its deficiencies.

Acknowledgments

This work was supported from CONACYT 254264 to F.G.G.

References

1. Burmester T, Weich B, Reinhardt S, Hankeln T. A vertebrate globin expressed in the brain. *Nature* 2000;**407**(6803):520–3. Sep 28 [cited 2018 Sep 24]. Available from: http://www.nature.com/articles/35035093.
2. Hankeln T, Ebner B, Fuchs C, Gerlach F, Haberkamp M, Laufs TL, et al. Neuroglobin and cytoglobin in search of their role in the vertebrate globin family. *J Inorg Biochem* 2005;**99**(1):110–19. Jan 1 [cited 2018 Sep 24]. Available from: https://www.sciencedirect.com/science/article/pii/S0162013404003514?via%3Dihub.
3. Kugelstadt D, Haberkamp M, Hankeln T, Burmester T. Neuroglobin, cytoglobin, and a novel, eye-specific globin from chicken. *Biochem Biophys Res Commun* 2004;**325**(3):719–25. Dec 17 [cited 2018 Sep 24]. Available from: http://linkinghub.elsevier.com/retrieve/pii/S0006291X04024106.

4. Roesner A, Fuchs C, Hankeln T, Burmester T. A globin gene of ancient evolutionary origin in lower vertebrates: evidence for two distinct globin families in animals. *Mol Biol Evol* 2005;**22**(1):12–20. Jan [cited 2018 Sep 24]. Available from: https://academic.oup.com/mbe/article-lookup/doi/10.1093/molbev/msh258.
5. Xie L-K, Yang S-H. Brain globins in physiology and pathology. *Med Gas Res* 2016;**6**(3):154–63. [cited 2018 Sep 24]. Available from: http://www.medgasres.com/text.asp?2016/6/3/154/191361.
6. Qiu XY, Chen XQ. Neuroglobin – recent developments. *Biomol Concepts* 2014;**5**(3):195–208. Jun 17 [cited 2018 Sep 24]. Available from: https://www.degruyter.com/view/j/bmc.ahead-of-print/bmc-2014-0011/bmc-2014-0011.xml.
7. Ascenzi P, di Masi A, Leboffe L, Fiocchetti M, Nuzzo MT, Brunori M, et al. Neuroglobin: from structure to function in health and disease. *Mol Aspects Med* 2016;**52**:1–48. Dec [cited 2018 Sep 24]. Available from: http://www.ncbi.nlm.nih.gov/pubmed/27825818.
8. Burmester T, Hankeln T. Neuroglobin: a respiratory protein of the nervous system. *News Physiol Sci* 2004;**19**:110–13. Jun [cited 2018 Sep 24]. Available from: http://www.ncbi.nlm.nih.gov/pubmed/15143204.
9. Pesce A, Dewilde S, Nardini M, Moens L, Ascenzi P, Hankeln T, et al. Human brain neuroglobin three-dimensional structure. *Structure* 2003;**11**:1087. Available from: http://www.rcsb.org/structure/1OJ6.
10. Alekseeva OS, Grigor'ev IP, Korzhevskii DE. Neuroglobin, an oxygen-binding protein in the mammalian nervous system (localization and putative functions). *J Evol Biochem Physiol* 2017;**53**(4):249–58. Jul 30 [cited 2018 Sep 24]. Available from: http://link.springer.com/10.1134/S0022093017040019.
11. Fabrizius A, Andre D, Laufs T, Bicker A, Reuss S, Porto E, et al. Critical re-evaluation of neuroglobin expression reveals conserved patterns among mammals. *Neuroscience* 2016;**337**:339–54. Nov 19 [cited 2018 Sep 24]. Available from: https://linkinghub.elsevier.com/retrieve/pii/S0306452216303578.
12. Hundahl CA, Kelsen J, Hay-Schmidt A. Neuroglobin and cytoglobin expression in the human brain. *Brain Struct Funct* 2013;**218**(2):603–9. Mar 17 [cited 2018 Sep 24]. Available from: http://link.springer.com/10.1007/s00429-012-0480-8.
13. Gilerovich EG, Grigor'ev IP, Kirik OV, Alekseeva OS, Sukhorukova EG, Korzhevskiĭ DÉ. Neuroglobin distribution in the human cerebellar cortex (an immunohistochemical study). *Morfologiia* 2014;**146**(4):75–7. [cited 2018 Sep 24]. Available from: http://www.ncbi.nlm.nih.gov/pubmed/25552092.
14. Kamioka Y, Fujikawa C, Ogai K, Sugitani K, Watanabe S, Kato S, et al. Functional characterization of fish neuroglobin: zebrafish neuroglobin is highly expressed in amacrine cells after optic nerve injury and can translocate into ZF4 cells. *Biochim Biophys Acta.* 2013;**1834**(9):1779–88. Sep [cited 2018 Sep 24]. Available from: http://linkinghub.elsevier.com/retrieve/pii/S1570963913000848.
15. Watanabe S, Takahashi N, Uchida H, Wakasugi K. Human neuroglobin functions as an oxidative stress-responsive sensor for neuroprotection. *J Biol Chem* 2012;**287**(36):30128–38. Aug 31 [cited 2018 Sep 24]. Available from: http://www.jbc.org/lookup/doi/10.1074/jbc.M112.373381.
16. Ostojić J, Grozdanić S, Syed NA, Hargrove MS, Trent III JT, et al. Neuroglobin and cytoglobin distribution in the anterior eye segment: a comparative immunohistochemical study. *J Histochem Cytochem* 2008;**56**(9):863. Sep [cited 2018 Sep 24]. Available from: http://www.ncbi.nlm.nih.gov/pubmed/18574250.
17. Nowotny M, Kiefer L, Andre D, Fabrizius A, Hankeln T, Reuss S. Hearing without neuroglobin. *Neuroscience* 2017;**366**:138–48. Dec 16 [cited 2018 Sep 24]. Available from: http://www.ncbi.nlm.nih.gov/pubmed/29031607.

18. Ilmjärv S, Reimets R, Hundahl CA, Luuk H. Effect of light on global gene expression in the neuroglobin-deficient mouse retina*Biomed Rep* 2014;**2**(6):780–6. Nov [cited 2018 Sep 24]. Available from: https://www.spandidos-publications.com/10.3892/br.2014.364.
19. Hundahl CA, Luuk H, Ilmjärv S, Falktoft B, Raida Z, Vikesaa J, et al. Neuroglobin-deficiency exacerbates Hif1A and c-FOS response, but does not affect neuronal survival during severe hypoxia in vivo. *PLoS One* 2011;**6**(12):e28160. [cited 2018 Sep 24]. Available from: http://www.ncbi.nlm.nih.gov/pubmed/22164238.
20. Picotti P, Dewilde S, Fago A, Hundahl C, De Filippis V, Moens L, et al. Unusual stability of human neuroglobin at low pH – molecular mechanisms and biological significance*FEBS J* 2009;**276**(23):7027–39. Dec 1 [cited 2018 Sep 24]. Available from: http://doi.wiley.com/10.1111/j.1742-4658.2009.07416.x.
21. Sun Y, Jin K, Mao XO, Zhu Y, Greenberg DA. Neuroglobin is up-regulated by and protects neurons from hypoxic-ischemic injury. *Proc Natl Acad Sci* 2001;**98**(26):15306–11. Dec 18 [cited 2018 Sep 24]. Available from: http://www.ncbi.nlm.nih.gov/pubmed/11742077.
22. Fordel E, Thijs L, Martinet W, Schrijvers D, Moens L, Dewilde S. Anoxia or oxygen and glucose deprivation in SH-SY5Y cells: a step closer to the unraveling of neuroglobin and cytoglobin functions*Gene* 2007;**398**(1–2):114–22. Aug 15 [cited 2018 Sep 24]. Available from: http://www.ncbi.nlm.nih.gov/pubmed/17532579.
23. Li RC, Lee SK, Pouranfar F, Brittian KR, Clair HB, Row BW, et al. Hypoxia differentially regulates the expression of neuroglobin and cytoglobin in rat brain. *Brain Res* 2006;**1096**(1):173–9. Jun 22 [cited 2018 Sep 24]. Available from: http://linkinghub.elsevier.com/retrieve/pii/S0006899306011152.
24. Liu Y, Li B, Li Q, Zou L. Neuroglobin up-regulation after ischaemic pre-conditioning in a rat model of middle cerebral artery occlusion. *Brain Inj* 2015;**29**(5):651–7. Apr 16 [cited 2018 Sep 24]. Available from: http://www.tandfonline.com/doi/full/10.3109/02699052.2014.1002004.
25. Shi S-Y, Feng X-M, Li Y, Li X, Chen X-L. Expression of neuroglobin in ocular hypertension induced acute hypoxic-ischemic retinal injury in rats. *Int J Ophthalmol* 2011;**4**(4):393–5. [cited 2018 Sep 24]. Available from: http://www.ncbi.nlm.nih.gov/pubmed/22553688.
26. Jin K, Mao Y, Mao X, Xie L, Greenberg DA. Neuroglobin expression in ischemic stroke. *Stroke* 2010;**41**(3):557–9. Mar [cited 2018 Sep 24]. Available from: http://www.ncbi.nlm.nih.gov/pubmed/20075359.
27. Hundahl C, Stoltenberg M, Fago A, Weber RE, Dewilde S, Fordel E, et al. Effects of short-term hypoxia on neuroglobin levels and localization in mouse brain tissues. *Neuropathol Appl Neurobiol* 2005;**31**(6):610–17. Dec [cited 2018 Sep 24]. Available from: http://www.ncbi.nlm.nih.gov/pubmed/16281909.
28. Mammen PPA, Shelton JM, Goetsch SC, Williams SC, Richardson JA, Garry MG, et al. Neuroglobin, a novel member of the globin family, is expressed in focal regions of the brain*J Histochem Cytochem* 2002;**50**(12):1591–8. Dec 26 [cited 2018 Sep 24]. Available from: http://www.ncbi.nlm.nih.gov/pubmed/12486081.
29. Fordel E, Geuens E, Dewilde S, De Coen W, Moens L. Hypoxia/ischemia and the regulation of neuroglobin and cytoglobin expression*IUBMB Life* 2004;**56**(11):681–7. Nov [cited 2018 Sep 24]. Available from: http://www.ncbi.nlm.nih.gov/pubmed/15804832.
30. Hümmler N, Schneider C, Giessl A, Bauer R, Walkinshaw G, Gassmann M, et al. Acute hypoxia modifies regulation of neuroglobin in the neonatal mouse brain. *Exp Neurol* 2012;**236**(1):112–21. Jul [cited 2018 Sep 24]. Available from: http://www.ncbi.nlm.nih.gov/pubmed/22548980.

31. Haines B, Demaria M, Mao X, Xie L, Campisi J, Jin K, et al. Hypoxia-inducible factor-1 and neuroglobin expression*Neurosci Lett* 2012;**514**(2):137–40. Apr 18 [cited 2018 Sep 24]. Available from: http://www.ncbi.nlm.nih.gov/pubmed/22342914.
32. Pinto-Almazán R, Segura-Uribe JJ, Farfán-García ED, Guerra-Araiza C. Effects of tibolone on the central nervous system: clinical and experimental approaches. *Biomed Res Int* 2017;**2017**:8630764. [cited 2018 Sep 24]. Available from: https://www.hindawi.com/journals/bmri/2017/8630764/.
33. Avila-Rodriguez M, Garcia-Segura LM, Hidalgo-Lanussa O, Baez E, Gonzalez J, Barreto GE. Tibolone protects astrocytic cells from glucose deprivation through a mechanism involving estrogen receptor beta and the upregulation of neuroglobin expression*Mol Cell Endocrinol* 2016;**433**:35–46. Sep 15 [cited 2018 Sep 25]. Available from: https://linkinghub.elsevier.com/retrieve/pii/S0303720716301897.
34. Wystub S, Ebner B, Fuchs C, Weich B, Burmester T, Hankeln T. Interspecies comparison of neuroglobin, cytoglobin and myoglobin: sequence evolution and candidate regulatory elements*Cytogenet Genome Res* 2004;**105**(1):65–78. [cited 2018 Sep 25]. Available from: https://www.karger.com/Article/FullText/78011.
35. Woo SK, Kwon MS, Geng Z, Chen Z, Ivanov A, Bhatta S, et al. Sequential activation of hypoxia-inducible factor 1 and specificity protein 1 is required for hypoxia-induced transcriptional stimulation of Abcc8. *J Cereb Blood Flow Metab* 2012;**32**(3):525–36. Mar 16 [cited 2018 Sep 25]. Available from: http://journals.sagepub.com/doi/10.1038/jcbfm.2011.159.
36. Hota KB, Hota SK, Srivastava RB, Singh SB. Neuroglobin regulates hypoxic response of neuronal cells through Hif-1α- and Nrf2-mediated mechanism. *J Cereb Blood Flow Metab* 2012;**32**(6):1046–60. Jun 4 [cited 2018 Sep 25] Available from: http://journals.sagepub.com/doi/10.1038/jcbfm.2012.21.
37. Sharp FR, Bernaudin M. HIF1 and oxygen sensing in the brain. *Nat Rev Neurosci* 2004;**5**(6):437–48. Jun 1 [cited 2018 Sep 25]. Available from: http://www.nature.com/articles/nrn1408.
38. Gao Y, Mengana Y, Cruz YR, Muñoz A, Testé IS, García JD, et al. Different expression patterns of Ngb and EPOR in the cerebral cortex and hippocampus revealed distinctive therapeutic effects of intranasal delivery of Neuro-EPO for ischemic insults to the gerbil brain*J Histochem Cytochem* 2011;**59**(2):214–27. Feb [cited 2018 Sep 25]. Available from: http://journals.sagepub.com/doi/10.1369/0022155410390323.
39. Jin K, Mao X, Xie L, Greenberg DA. Interactions between vascular endothelial growth factor and neuroglobin*Neurosci Lett* 2012;**519**(1):47–50. Jun 21 [cited 2018 Sep 25]. Available from: http://linkinghub.elsevier.com/retrieve/pii/S0304394012006544.
40. Yu Z, Xu J, Liu N, Wang Y, Li X, Pallast S, et al. Mitochondrial distribution of neuroglobin and its response to oxygen-glucose deprivation in primary-cultured mouse cortical neurons*Neuroscience* 2012;**218**:235–42. Aug 30 [cited 2018 Sep 25]. Available from: http://linkinghub.elsevier.com/retrieve/pii/S0306452212005453.
41. Lechauve C, Augustin S, Cwerman-Thibault H, Bouaita A, Forster V, Célier C, et al. Neuroglobin involvement in respiratory chain function and retinal ganglion cell integrity*Biochim Biophys Acta* 2012;**1823**(12):2261–73. Dec [cited 2018 Sep 25]. Available from: http://linkinghub.elsevier.com/retrieve/pii/S0167488912002789.
42. Bentmann A, Schmidt M, Reuss S, Wolfrum U, Hankeln T, Burmester T. Divergent distribution in vascular and avascular mammalian retinae links neuroglobin to cellular respiration*J Biol Chem* 2005;**280**(21):20660–5. May 27 [cited 2018 Sep 25]. Available from: http://www.jbc.org/lookup/doi/10.1074/jbc.M501338200.
43. Mitz SA, Reuss S, Folkow LP, Blix AS, Ramirez J-M, Hankeln T, et al. When the brain goes diving: glial oxidative metabolism may confer hypoxia tolerance to the seal brain*Neuroscience* 2009;**163**(2):552–60. Oct 6 [cited 2018 Sep 25]. Available from: http://linkinghub.elsevier.com/retrieve/pii/S0306452209011075.

44. Yu Z, Zhang Y, Liu N, Yuan J, Lin L, Zhuge Q, et al. Roles of neuroglobin binding to mitochondrial complex III subunit cytochrome c1 in oxygen-glucose deprivation-induced neurotoxicity in primary neurons*Mol Neurobiol* 2016;**53**(5):3249–57. Jul 7 [cited 2018 Sep 25]. Available from: http://link.springer.com/10.1007/s12035-015-9273-4.
45. Fordel E, Thijs L, Moens L, Dewilde S. Neuroglobin and cytoglobin expression in mice. Evidence for a correlation with reactive oxygen species scavenging. *FEBS J* 2007;**274**(5):1312–17. Mar [cited 2018 Sep 25]. Available from: http://doi.wiley.com/10.1111/j.1742-4658.2007.05679.x.
46. Nayak G, Prentice HM, Milton SL. Role of neuroglobin in regulating reactive oxygen species in the brain of the anoxia-tolerant turtle *Trachemys scripta*. *J Neurochem* 2009;**110**(2):603–12. Jul [cited 2018 Sep 25]. Available from: http://doi.wiley.com/10.1111/j.1471-4159.2009.06157.x.
47. Giuffrè A, Stubauer G, Sarti P, Brunori M, Zumft WG, Buse G, et al. The heme-copper oxidases of *Thermus thermophilus* catalyze the reduction of nitric oxide: evolutionary implications*Proc Natl Acad Sci USA* 1999;**96**(26):14718–23. Dec 21 [cited 2018 Sep 25]. Available from: http://www.ncbi.nlm.nih.gov/pubmed/10611279.
48. Liu J, Yu Z, Guo S, Lee S-R, Xing C, Zhang C, et al. Effects of neuroglobin overexpression on mitochondrial function and oxidative stress following hypoxia/reoxygenation in cultured neurons*J Neurosci Res* 2009;**87**(1):164–70. Jan [cited 2018 Sep 25]. Available from: http://doi.wiley.com/10.1002/jnr.21826.
49. De Marinis E, Fiocchetti M, Acconcia F, Ascenzi P, Marino M. Neuroglobin upregulation induced by 17β-estradiol sequesters cytochrome c in the mitochondria preventing H2O2-induced apoptosis of neuroblastoma cells. *Cell Death Dis* 2013;**4**(2):e508. Feb 21 [cited 2018 Sep 25]. Available from: http://www.nature.com/articles/cddis201330.
50. Guidolin D, Agnati LF, Tortorella C, Marcoli M, Maura G, Albertin G, et al. Neuroglobin as a regulator of mitochondrial-dependent apoptosis: a bioinformatics analysis. *Int J Mol Med* 2014;**33**(1):111–16. Jan [cited 2018 Sep 25]. Available from: https://www.spandidos-publications.com/10.3892/ijmm.2013.1564.
51. Yu Z, Liu N, Li Y, Xu J, Wang X. Neuroglobin overexpression inhibits oxygen-glucose deprivation-induced mitochondrial permeability transition pore opening in primary cultured mouse cortical neurons*Neurobiol Dis* 2013;**56**:95–103. Aug [cited 2018 Sep 25]. Available from: http://linkinghub.elsevier.com/retrieve/pii/S0969996113001290.
52. Lan W-B, Lin J-H, Chen X-W, Wu C-Y, Zhong G-X, Zhang L-Q, et al. Overexpressing neuroglobin improves functional recovery by inhibiting neuronal apoptosis after spinal cord injury*Brain Res* 2014;**1562**:100–8. May [cited 2018 Sep 25]. Available from: http://linkinghub.elsevier.com/retrieve/pii/S0006899314003680.
53. Madesh M, Hajnóczky G. VDAC-dependent permeabilization of the outer mitochondrial membrane by superoxide induces rapid and massive cytochrome c release. *J Cell Biol* 2001;**155**(6):1003–15. Dec 10 [cited 2018 Sep 25]. Available from: http://www.jcb.org/lookup/doi/10.1083/jcb.200105057.
54. Khan AA, Mao XO, Banwait S, DerMardirossian CM, Bokoch GM, Jin K, et al. Regulation of hypoxic neuronal death signaling by neuroglobin. *FASEB J* 2008;**22**(6):1737–47. Jun [cited 2018 Sep 25]. Available from: http://www.ncbi.nlm.nih.gov/pubmed/18198211.
55. Lin Y, Cai B, Xue X-H, Fang L, Wu Z-Y, Wang N. TAT-mediated delivery of neuroglobin attenuates apoptosis induced by oxygen-glucose deprivation via the Jak2/Stat3 pathway in vitro*Neurol Res* 2015;**37**(6):531–8. Jun 15 [cited 2018 Sep 25]. Available from: http://www.tandfonline.com/doi/full/10.1179/1743132814Y.0000000420.
56. Brittain T, Skommer J, Henty K, Birch N, Raychaudhuri S. A role for human neuroglobin in apoptosis*IUBMB Life* 2010;**62**(12):878–85. Dec [cited 2018 Sep 25]. Available from: http://www.ncbi.nlm.nih.gov/pubmed/21190290.

57. Raychaudhuri S, Skommer J, Henty K, Birch N, Brittain T. Neuroglobin protects nerve cells from apoptosis by inhibiting the intrinsic pathway of cell death. *Apoptosis* 2010;**15**(4):401–11. Apr 20 [cited 2018 Sep 25]. Available from: http://link.springer.com/10.1007/s10495-009-0436-5.
58. Brittain T. The anti-apoptotic role of neuroglobin. *Cells* 2012;**1**(4):1133–55. Nov 23 [cited 2018 Sep 25]. Available from: http://www.ncbi.nlm.nih.gov/pubmed/24710547.
59. Wakasugi K, Nakano T, Morishima I. Oxidized human neuroglobin acts as a heterotrimeric Galpha protein guanine nucleotide dissociation inhibitor. *J Biol Chem* 2003;**278**(38):36505–12. Sep 19 [cited 2018 Sep 25]. Available from: http://www.jbc.org/lookup/doi/10.1074/jbc.M305519200.
60. Antao ST, Duong TTH, Aran R, Witting PK. Neuroglobin overexpression in cultured human neuronal cells protects against hydrogen peroxide insult via activating phosphoinositide-3 kinase and opening the mitochondrial K(ATP) channel. *Antioxid Redox Signal* 2010;**13**(6):769–81. Sep 15 [cited 2018 Sep 25]. Available from: http://www.liebertpub.com/doi/10.1089/ars.2009.2977.
61. Li Y, Dai Y, Sun J, Xiang Y, Yang J, Dai S, et al. Neuroglobin attenuates beta amyloid-induced apoptosis through inhibiting caspases activity by activating PI3K/Akt signaling pathway*J Mol Neurosci* 2016;**58**(1):28–38. Jan 7 [cited 2018 Sep 25]. Available from: http://link.springer.com/10.1007/s12031-015-0645-z.
62. Cai B, Li W, Mao X, Winters A, Ryou M-G, Liu R, et al. Neuroglobin overexpression inhibits AMPK signaling and promotes cell anabolism. *Mol Neurobiol* 2016;**53**(2):1254–65. Mar 24 [cited 2018 Sep 25]. Available from: http://link.springer.com/10.1007/s12035-014-9077-y.
63. Liu N, Yu Z, Li Y, Yuan J, Zhang J, Xiang S, et al. Transcriptional regulation of mouse neuroglobin gene by cyclic AMP responsive element binding protein (CREB) in N2a cells. *Neurosci Lett* 2013;**534**:333–7. Feb 8 [cited 2018 Sep 25]. Available from: http://linkinghub.elsevier.com/retrieve/pii/S0304394012014826.
64. Sun F, Mao X, Xie L, Greenberg DA, Jin K. Neuroglobin protein is upregulated in Alzheimer's Disease*J Alzheimer's Dis* 2013;**36**(4):659–63. Aug 5 [cited 2018 Sep 25]. Available from: http://www.ncbi.nlm.nih.gov/pubmed/23648513.
65. Szymanski M, Wang R, Fallin MD, Bassett SS, Avramopoulos D. Neuroglobin and Alzheimer's dementia: genetic association and gene expression changes. *Neurobiol Aging* 2010;**31**(11):1835–42. Nov [cited 2018 Sep 25]. Available from: http://linkinghub.elsevier.com/retrieve/pii/S0197458008003679.
66. Chen L-M, Xiong Y-S, Kong F-L, Qu M, Wang Q, Chen X-Q, et al. Neuroglobin attenuates Alzheimer-like tau hyperphosphorylation by activating Akt signaling. *J Neurochem* 2012;**120**(1):157–64. Jan [cited 2018 Sep 25]. Available from: http://doi.wiley.com/10.1111/j.1471-4159.2011.07275.x.
67. Lu T, Aron L, Zullo J, Pan Y, Kim H, Chen Y, et al. REST and stress resistance in ageing and Alzheimer's disease*Nature* 2014;**507**(7493):448–54. Mar 19 [cited 2018 Sep 25]. Available from: http://www.ncbi.nlm.nih.gov/pubmed/24670762.
68. Zhang W, Tian Z, Sha S, Cheng LYL, Philipsen S, Tan-Un K-C. Functional and sequence analysis of human neuroglobin gene promoter region. *Biochim Biophys Acta* 2011;**1809**(4–6):236–44. Apr [cited 2018 Sep 25]. Available from: http://linkinghub.elsevier.com/retrieve/pii/S1874939911000460.
69. Van Acker ZP, Luyckx E, Van Leuven W, Geuens E, De Deyn PP, Van Dam D, et al. Impaired hypoxic tolerance in APP23 mice: a dysregulation of neuroprotective globin levels*FEBS Lett* 2017;**591**(10):1321–32. May [cited 2018 Sep 25]. Available from: http://doi.wiley.com/10.1002/1873-3468.12651.
70. Zara S, De Colli M, Rapino M, Pacella S, Nasuti C, Sozio P, et al. Ibuprofen and lipoic acid conjugate neuroprotective activity is mediated by Ngb/Akt intracellular signaling pathway in Alzheimer's disease rat model*Gerontology* 2013;**59**(3):250–60. [cited 2018 Sep 25]. Available from: https://www.karger.com/Article/FullText/346445.

71. Jin K, Mao XO, Xie L, John V, Greenberg DA. Pharmacological induction of neuroglobin expression*Pharmacology* 2011;**87**(1–2):81–4. [cited 2018 Sep 25]. Available from: http://www.ncbi.nlm.nih.gov/pubmed/21228614.
72. Di Pietro V, Lazzarino G, Amorini AM, Tavazzi B, D'Urso S, Longo S, et al. Neuroglobin expression and oxidant/antioxidant balance after graded traumatic brain injury in the rat*Free Radic Biol Med* 2014;**69**:258–64. Apr [cited 2018 Sep 25]. Available from: http://linkinghub.elsevier.com/retrieve/pii/S0891584914000458.
73. Purushothuman S, Stone J. The reaction of cerebral cortex to a nearby lesion: damage, survival, self-protection*Brain Res* 2015;**1601**:52–63. Mar 19 [cited 2018 Sep 25]. Available from: http://linkinghub.elsevier.com/retrieve/pii/S0006899315000062.
74. Sun Y, Jin K, Mao XO, Xie L, Peel A, Childs JT, et al. Effect of aging on neuroglobin expression in rodent brain. *Neurobiol Aging* 2005;**26**(2):275–8. Feb [cited 2018 Sep 25]. Available from: http://linkinghub.elsevier.com/retrieve/pii/S0197458004001095.
75. Fiocchetti M, De Marinis E, Ascenzi P, Marino M. Neuroglobin and neuronal cell survival*Biochim Biophys Acta* 2013;**1834**(9):1744–9. Sep [cited 2018 Sep 25]. Available from: http://www.ncbi.nlm.nih.gov/pubmed/23357651.
76. Hundahl CA, Allen GC, Hannibal J, Kjaer K, Rehfeld JF, Dewilde S, et al. Anatomical characterization of cytoglobin and neuroglobin mRNA and protein expression in the mouse brain*Brain Res* 2010;**1331**:58–73. May 17 [cited 2018 Sep 25]. Available from: http://linkinghub.elsevier.com/retrieve/pii/S0006899310006293.
77. Hundahl CA, Allen GC, Nyengaard JR, Dewilde S, Carter BD, Kelsen J, et al. Neuroglobin in the rat brain: localization*Neuroendocrinology* 2008;**88**(3):173–82. [cited 2018 Sep 25]. Available from: http://www.ncbi.nlm.nih.gov/pubmed/18451642.
78. Szymusiak R, Alam N, McGinty D. Discharge patterns of neurons in cholinergic regions of the basal forebrain during waking and sleep. *Behav Brain Res* 2000;**115**(2):171–82. Nov [cited 2018 Sep 25]. Available from: http://www.ncbi.nlm.nih.gov/pubmed/11000419.
79. Melgarejo-Gutiérrez M, Acosta-Peña E, Venebra-Muñoz A, Escobar C, Santiago-García J, Garcia-Garcia F. Sleep deprivation reduces neuroglobin immunoreactivity in the rat brain. *Neuroreport* 2013;**24**(3):120–5. Feb 13 [cited 2018 Sep 25]. Available from: http://www.ncbi.nlm.nih.gov/pubmed/23262504.
80. Amri F, Ghouili I, Amri M, Carrier A, Masmoudi-Kouki O. Neuroglobin protects astroglial cells from hydrogen peroxide-induced oxidative stress and apoptotic cell death. *J Neurochem* 2017;**140**(1):151–69. Jan [cited 2018 Sep 27]. Available from: http://www.ncbi.nlm.nih.gov/pubmed/27787886.
81. Gopalakrishnan A, Ji LL, Cirelli C. Sleep deprivation and cellular responses to oxidative stress. *Sleep* 2004;**27**(1):27–35. Feb 1 [cited 2018 Sep 25]. Available from: http://www.ncbi.nlm.nih.gov/pubmed/14998234.
82. Everson CA, Henchen CJ, Szabo A, Hogg N. Cell injury and repair resulting from sleep loss and sleep recovery in laboratory rats*Sleep* 2014;**37**(12):1929–40. Dec 1 [cited 2018 Sep 25]. Available from: https://academic.oup.com/sleep/article-lookup/doi/10.5665/sleep.4244.
83. Peyron C, Tighe DK, van den Pol AN, de Lecea L, Heller HC, Sutcliffe JG, et al. Neurons containing hypocretin (orexin) project to multiple neuronal systems. *J Neurosci* 1998;**18**(23):9996–10015. Dec 1 [cited 2018 Sep 27]. Available from: http://www.ncbi.nlm.nih.gov/pubmed/9822755.
84. Hundahl CA, Kelsen J, Dewilde S, Hay-Schmidt A. Neuroglobin in the rat brain (II): co-localisation with neurotransmitters*Neuroendocrinology* 2008;**88**(3):183–98. [cited 2018 Sep 27]. Available from: http://www.ncbi.nlm.nih.gov/pubmed/18509243.
85. Xu M, Yang Y, Zhang J. Levels of neuroglobin in serum and neurocognitive impairments in Chinese patients with obstructive sleep apnea. *Sleep Breath* 2013;**17**(2):573–82. May 7 [cited 2018 Sep 25]. Available from: http://link.springer.com/10.1007/s11325-012-0723-1.

86. Nair D, Ramesh V, Gozal D. Cognitive deficits are attenuated in neuroglobin overexpressing mice exposed to a model of obstructive sleep apnea. *Front Neurol* 2018;**9**:426. Jun 5 [cited 2018 Sep 27]. Available from: https://www.frontiersin.org/article/10.3389/fneur.2018.00426/full.
87. Shang A, Zhou D, Wang L, Gao Y, Fan M, Wang X, et al. Increased neuroglobin levels in the cerebral cortex and serum after ischemia-reperfusion insults. *Brain Res* 2006;**1078**(1):219–26. Mar 17 [cited 2018 Sep 25]. Available from: http://linkinghub.elsevier.com/retrieve/pii/S0006899306001715.
88. Hundahl CA, Hannibal J, Fahrenkrug J, Dewilde S, Hay-Schmidt A. Neuroglobin expression in the rat suprachiasmatic nucleus: colocalization, innervation, and response to light*J Comp Neurol* 2010;**518**(9):1556–69. May 1 [cited 2018 Sep 25]. Available from: http://doi.wiley.com/10.1002/cne.22290.
89. Hundahl CA, Fahrenkrug J, Hay-Schmidt A, Georg B, Faltoft B, Hannibal J. Circadian behaviour in neuroglobin deficient mice*PLoS One* 2012;**7**(4):e34462. Apr 5 [cited 2018 Sep 25]. Available from: http://dx.plos.org/10.1371/journal.pone.0034462.

CHAPTER

6

Overview of Sleep Disorders

Aljohara S. Almeneessier[1,2], *Ravi Gupta*[3], *Seithikurippu R. Pandi-Perumal*[4] *and Ahmed S. BaHammam*[1]

[1]University Sleep Disorders Center, Department of Medicine, College of Medicine, King Saud University, Riyadh, Saudi Arabia [2]Department of Family Medicine, College of Medicine, King Saud University, Riyadh, Saudi Arabia [3]Department of Psychiatry, All India Institute of Medical Sciences, Rishikesh, India [4]Somnogen Canada Inc., Toronto, ON, Canada

OUTLINE

6.1 INTRODUCTION

Sleep is an important physiological function of the body, and sleep disturbances often lead to alterations in several other physiological functions, as well as increased morbidity and mortality. The new version of the *International Classification of Sleep Disorders*, 3rd Edition (ICSD-3), published in 2014 by the American Academy of Sleep Medicine (AASM), classifies sleep disorders and describes the

Molecular Sleep—Wake Cycle
DOI: https://doi.org/10.1016/B978-0-12-816430-3.00006-3

TABLE 6.1 ICSD-3 Major Diagnostic Sections[1]

Section
Insomnia
Sleep-related breathing disorders
Circadian rhythm sleep–wake disorders
Parasomnias
Sleep-related movement disorders
Other sleep disorders

diagnostic criteria. Diagnostic codes for each disorder are also provided. Currently, more than 60 sleep disorders are described in the ICSD-3 (Table 6.1).[2]

Although several disorders may cause sleep disturbances and impaired daytime function, sleep disorders are often underrecognized and undertreated in clinical practice. In this chapter, we discuss several common sleep disorders that are frequently encountered in daily practice, focusing on obstructive sleep apnea (OSA) and excessive daytime sleepiness (EDS).

6.1.1 Obstructive Sleep Apnea

6.1.1.1 Pathophysiology

The upper airway tends to narrow during sleep, resulting in recurrent closures, which may cause repeated apneas despite continued efforts to breathe. This, in turn, results in intermittent hypoxemia and frequent arousals. Intermittent hypoxemia and arousals cause an increase in sympathetic activity, thereby increasing the risk of cardiovascular complications and arrhythmia. Frequent apneas during sleep disturb sleep architecture, resulting in poor sleep quality and in some patients daytime sleepiness.

6.1.1.2 Prevalence

Obstructive sleep apnea syndrome (OSAS) is defined as five or more abnormal obstructed breathing events during sleep and excessive sleepiness during the day. Studies have reported that the prevalence of OSAS is 4% in men and 2% in women. More recent studies have shown that the prevalence of OSAS has increased to 7% in adult men and to 5% in adult women.[2] However, the prevalence of OSA without daytime sleepiness is much higher.[3] Nevertheless, OSA remains considerably underdiagnosed, with an estimated 75%–80% of cases remaining unidentified.[4] In addition, studies have suggested a delay between symptom onset and referral to sleep disorders centers in patients with OSA,[5–7] resulting in a significant delay in diagnosis and treatment. This delay may result in accumulated damage to the cardiovascular system.[8]

6.1.1.3 *Symptoms and Presentation*

Snoring is the most common symptom of OSA, although not all individuals who snore have OSA. Patients with OSA may also present with choking attacks during sleep, witnessed apnea, nocturia, mouth breathing with dryness of mouth and throat on awakening, excessive salivation during sleep, morning headache, excessive sweating, nocturnal heartburn, nocturnal palpitation, unrefreshing sleep, and EDS. However, it is important to note that only 20%–30% of patients with OSA present with EDS. Research has indicated that patients with OSA may also present with insomnia and that the rate of insomnia appears to be higher in women than men.[9] Obstructive events may occur mainly during rapid eye movement (REM) sleep in a subgroup of patients. REM-predominant OSA is observed mainly in women, children, and young adults. Usually, patients with REM-related OSA have milder symptoms; however, they may present with nightmares and vivid dreams.[10]

Several questionnaires have been developed to screen for OSA, the most common of which is the Berlin questionnaire and the STOP-Bang questionnaire. Both questionnaires have been validated in several languages.[11,12]

6.1.1.4 *Risk Factors for Obstructive Sleep Apnea*

Several risk factors have been associated with OSA, including the following:

- Obesity: Approximately 70% of patients with OSA are obese. Nevertheless, severe OSA can be observed in nonobese patients with craniofacial abnormalities.
- Increased neck circumference (>17 in. for men and >16 in. in women).
- Gender: OSA is more prevalent in men.
- Age: OSA is more prevalent in older people. The risk in women significantly increases postmenopause.
- Craniofacial abnormalities affecting the jaw size: retro- and micrognathia (appears as a small mandible and an overbite).
- Enlarged tonsils and adenoids, particularly in children; nevertheless, these can occasionally be observed in adults.

6.1.1.5 Complications of Obstructive Sleep Apnea

OSA can lead to EDS, which increases the risk of motor vehicle and other accidents. In addition, there is a strong link between OSA and cardiovascular and cerebrovascular complications such as hypertension, ischemic heart disease, arrhythmias, heart failure, pulmonary hypertension, and stroke. Recent studies have also linked OSA with insulin resistance, as well as depression, decreased memory, and poor concentration.[13]

6.1.1.6 Diagnosis

The AASM considers polysomnography (PSG) to be standard for the diagnosis of sleep-disordered breathing.[14] Neurological, cardiovascular, and respiratory parameters are monitored during sleep via PSG, an attended sleep study usually performed overnight in specialized settings. PSG allows for determination of total sleep time, sleep stages, arousals, snoring, apneas, hypopneas, effort-related arousals and desaturation. PSG also allows the distinction between obstructive and central sleep apnea. Fig. 6.1 depicts respiratory monitoring during polysomnography: (A) abdominal and thoracic belts to monitor thoraco-abdominal breathing and (B) thermistor and nasal cannula to monitor airflow. Every 30 seconds of a sleep study appears on a separate page and is referred to as an "epoch." A classical sleep study of 7 hours constitutes around 800 pages. Sleep studies are scored manually epoch by epoch. Fig. 6.2 shows one epoch of a patient with OSA.

Recently, unattended home sleep testing using portable sleep monitoring devices has become popular. The most commonly used portable sleep study devices are level-III devices, which allow for the monitoring of airflow, respiratory effort, oxygen saturation, and heart rate. Although more convenient for patients and less costly than traditional PSG, portable devices have several limitations including the inability to monitor EEG signals, which prevents the determination of sleep duration and sleep stages. Moreover, being unattended, this type of monitoring has a relatively higher rate of signal failure.

6.1.1.7 Definitions of Apnea and Hypopnea

Apnea is defined as the complete cessation (90% or more reduction of nasal or oral airflow) of breathing for a duration of 10 seconds or greater. In contrast, hypopnea is defined as at least 30% drop in nasal airflow, followed by an at least 4% drop in SaO_2. Fig. 6.3 demonstrates a classic example of an obstructive apnea during PSG monitoring.

6.1.1.8 Assessment of Obstructive Sleep Apnea Severity

The severity of OSA is determined by the number of apneas and hypopneas per hour of sleep and reported as an apnea–hypopnea

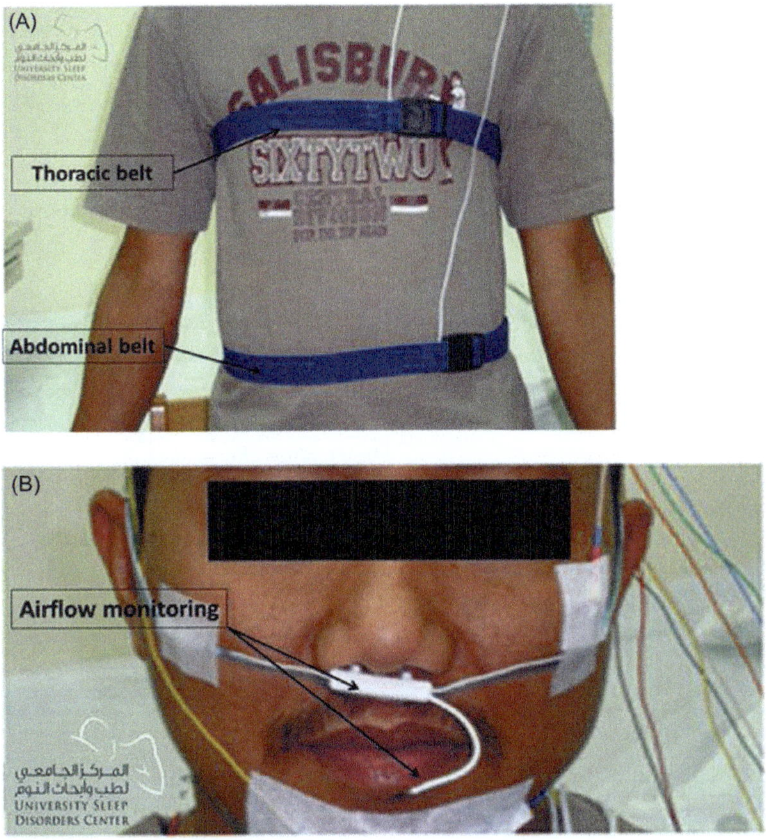

FIGURE 6.1 Respiratory monitoring during polysomnography. (A) Abdominal and thoracic belts are used to monitor thoraco-abdominal breathing. (B) A thermistor and nasal cannula are used to monitor airflow.

index (AHI): AHI <5/hour = normal; 5–15/hour = mild; 15–30/hour = moderate; and >30/hour = severe. Other parameters that may indicate the severity of OSA include the desaturation index (the number desaturations with a 3–4% drop in O_2 compared with baseline/hour of sleep) and time spent with O_2 saturation less than 90%.

6.1.1.9 Treatment

The mainstay of treatment of OSA is noninvasive positive airway pressure (PAP) therapy, which is applied via an interfacing mask in the form of continuous positive airway pressure (CPAP) or bilevel PAP. The AASM considers PAP therapy the treatment of choice for mild, moderate, and severe OSA.[15] PAP therapy results in significant improvement in symptoms and several studies have shown that PAP

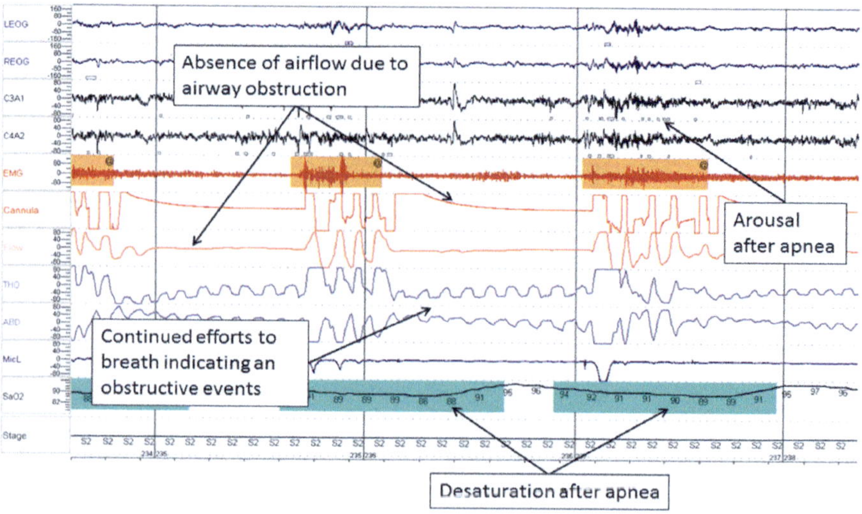

FIGURE 6.2 An example of a polysomnographic study of a patient with obstructive sleep apnea. LEOG and REOG: left and right electrooculogram; C3A1 and C4A2: EEG channels; EMG: chin electromyogram; Cannula and Flow: measurements of airflow; THO and ABD: thoracic and abdominal movements; SaO_2: oxygen saturation.

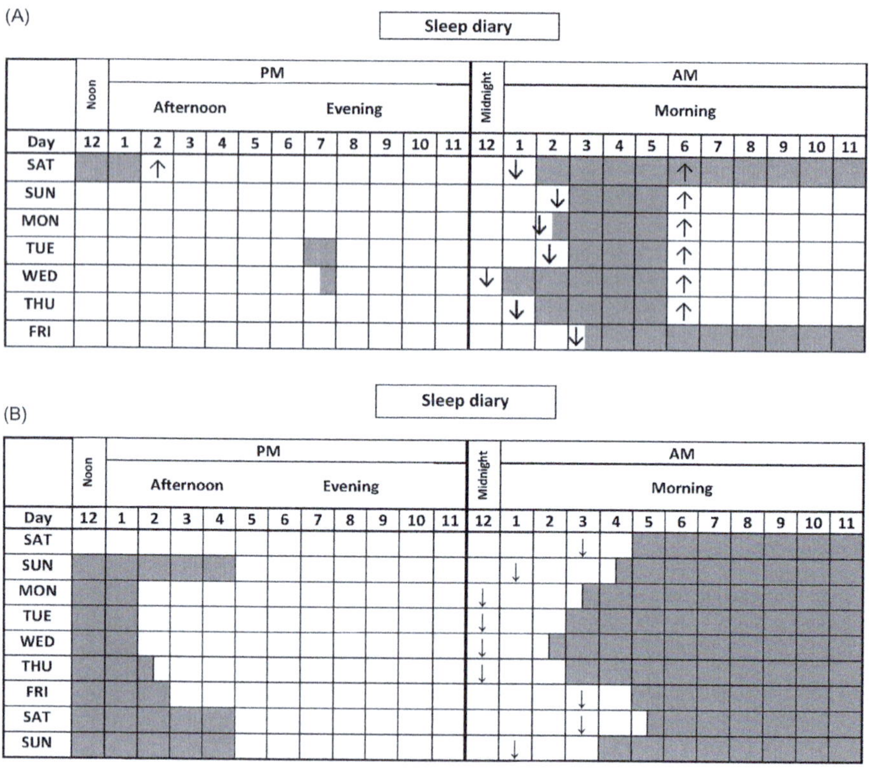

FIGURE 6.3 Sleep diary for common circadian rhythm sleep disorders. (A) Insufficient sleep syndrome. (B) Delayed sleep phase disorder.

therapy also reduces cardiovascular complications of OSA. Although the side effects of PAP therapy are usually minor and reversible (e.g., dry mouth and throat, eye irritation, stomach distension, claustrophobia, and nasal bridge irritation), adherence to PAP therapy remains a major obstacle. Therefore, patients should receive systematic education to improve adherence, and PAP usage should be monitored objectively.[16]

6.1.1.10 PAP Therapy Titration

Ideally, CPAP titration should be performed manually under PSG monitoring. The best fitting mask that is accepted by the patient should be used. Under PSG monitoring, CPAP pressure is increased gradually as per a written protocol until snoring, arousals, and all obstructive respiratory events have been eliminated.

6.1.1.11 Alternative Treatment for Obstructive Sleep Apnea

Other treatment modalities have been proposed for patients with OSA; however, the evidence supporting their efficacy is not as clear as that for PAP therapy. In this section, we summarize the most important alternative treatment modalities for OSA:

- *Mandibular advancement devices (MADs)*:

  MADs are designed by specialized orthodontists to reposition the jaw forward and downward during sleep to widen the upper airway and improve its patency. Current evidence supports the use of MADs in patients with mild to moderate OSA who do not tolerate PAP therapy.[17]
- *Surgery*:

  Several surgical procedures have been used to treat OSA; however, the evidence supporting most of those modalities is weak.
  - *Tonsillectomy and adenotonsillectomy*:

    Adenotonsillectomy is recommended for the treatment of OSA in children with adenotonsillar hypertrophy. In adults, tonsillectomy is indicated for the treatment of OSA in the presence of tonsillar hypertrophy.[17]
  - *Uvulopalatopharyngoplasty (UPPP)*:

    UPPP aims to reduce upper airway obstruction at the oropharyngeal level by reducing soft palate redundancy. In UPPP, the retropalatal airway is expanded by trimming the posterior and anterior lateral pillars and excising the uvula and posterior soft palate.[17] However, randomized clinical trials comparing the efficacy of UPPP with no treatment or sham surgery in patients with OSA are lacking. Overall, studies have reported a success rate of 40%–50% after UPPP in patients with mild to moderate OSA.

Nevertheless, the efficacy of UPPP appears to diminish over time. Failure of UPPP is usually attributed to multiple sites of obstruction located more caudally in the upper airway, or to the persistence of the retropalatal obstruction. Based on the available evidence, UPPP cannot be recommended except in carefully selected patients.[17]

- *Nasal surgery*:

The rationale behind this type of surgery is that improving nasal airway patency and reducing nasal resistance may alleviate sleep-disordered breathing. Nevertheless, based on the available data, nasal surgery as a single intervention is not recommended for the treatment of OSA.[17]

6.1.2 Excessive Daytime Sleepiness (Hypersomnolence)

6.1.2.1 Definition of Excessive Daytime Sleepiness

Daytime sleepiness can be defined as the inability to maintain wakefulness during waking hours, resulting in unplanned periods of sleep. It is important to note that sleepiness and fatigue have different meanings.[1] Daytime sleepiness has been investigated in several studies throughout the world, with a reported prevalence ranging from 9% to 26%.[18] Cultural differences in patterns of sleep, genetic variation, work schedules, and social and cultural activities in different countries may play a role in the level of sleepiness.

Sleepiness is a subjective perception of sleep tendency, especially in low stimulus conditions, which reflects a biological need to sleep. The importance of sleep to sleepiness can be viewed as the importance of food to hunger. In contrast, fatigue is a physical or psychological feeling that refers to many different conditions, some of which do not include sleepiness. Fatigue refers specifically to increased difficulty sustaining a high level of performance. Fatigue may accompany a variety of disorders, such as depression or neurodegenerative diseases.[19,20] Unlike sleepiness, patients with fatigue do not fall asleep in low stimulus conditions, such as while sitting in waiting areas or reading. This distinction is important because sleepiness indicates lack of sleep, the presence of a sleep disorder that disturbs sleep, or a central hypersomnia disorder.

EDS has serious consequences, including increased risk of injury, car accidents, and lower productivity.[21] The most common causes of EDS are insufficient sleep syndrome (ISS), shift work disorder, and OSA. Central causes of EDS, such as narcolepsy and idiopathic hypersomnia, are less prevalent and therefore underrecognized. Table 6.2 includes some of the causes of EDS.

TABLE 6.2 Causes of Excessive Daytime Sleepiness

1.	Hypersomnia due to medical or psychiatric disorder or drug or substance
2.	Narcolepsy type 1 or 2
3.	Idiopathic hypersomnia
4.	Kleine–Levin syndrome (periodic hypersomnia) (a very rare disorder)
5.	Insufficient sleep syndrome (sleep deprivation)
6.	Obstructive sleep apnea
7.	Central sleep apnea
8.	Shift work disorder
9.	Delayed sleep phase disorder
10.	Advanced sleep phase disorder
11.	Long sleeper
12.	Periodic limb movement disorder
13.	Hypersomnia due to a medical disorder
14.	Hypersomnia due to a medication or substance

6.1.2.2 Assessment of Excessive Daytime Sleepiness

A systematic approach is essential to distinguish between the different causes of EDS. The cause of EDS can often be suspected based on a patient's history. Therefore, a detailed sleep history should be obtained from the patient and the patient's bed partner or caregiver. History should include assessment of bedtime, wake time, and sleep duration. Medical, psychiatric, and surgical history should also be evaluated to ensure a comprehensive differential diagnosis, along with a family history and current medications. This should be followed by a focused physical exam to narrow the differential diagnosis.

Several questionnaires have been utilized to assess daytime sleepiness, the most common of which is the Epworth Sleepiness Scale (ESS). The ESS is a validated eight-item questionnaire that assesses the likelihood that the individual will fall asleep during certain activities.[22] The questionnaire asks the individual to rate his or her probability of falling asleep on a scale of increasing probability from 0 to 3 for eight common situations in daily life. Total scores range from 0 to 24, and a score of ≥ 10 indicates increased daytime sleepiness.

When a patient presents with EDS, it is essential to ensure that the patient is not sleep deprived. Therefore, a 2-week sleep diary should be obtained to determine if there is sufficient sleep duration per night or whether a circadian rhythm disorder, such as an advanced or delayed sleep phase disorder, is present (see Fig. 6.3). Fig. 6.4 shows a proposed algorithm for the assessment of patients presenting with EDS.

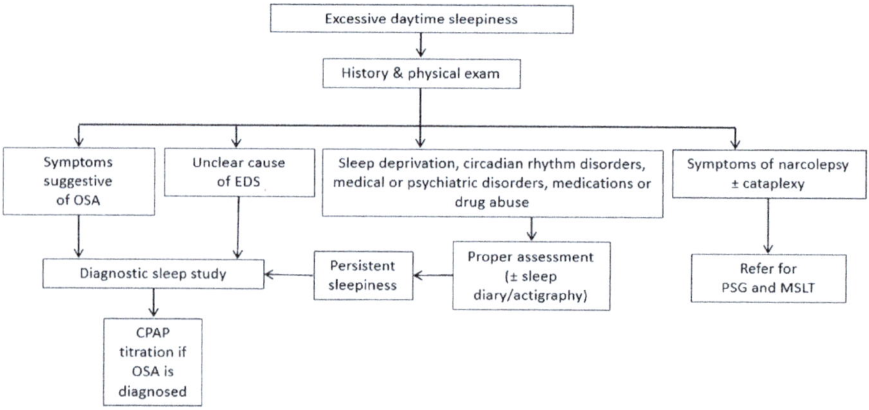

FIGURE 6.4 A proposed algorithm for the assessment of patients with excessive sleepiness.

6.1.2.3 Causes of Excessive Daytime Sleepiness

6.1.2.3.1 Insufficient Sleep Syndrome (ISS)

Nocturnal sleep duration of 7–8 hours is thought to be required to maintain good health in adults.[23] A study investigating the ideal duration of sleep to prevent deficits in psychomotor vigilance found that an average of 8.2 hours is needed per night for best performance.[24] However, the US Department of Health and Human Services estimates that only 65% of Americans get a healthy amount of sleep.[25] In the third edition of the *International Classification of Sleep Disorders* (ICSD-3),[1] ISS was classified as a central hypersomnolence disorder characterized by longer sleep durations on weekends than on weekdays, as well as shorter sleep times during weekdays and daytime sleepiness. Sleep diaries and actigraphy are often used to assess for ISS. Fig. 6.3A shows the sleep diary of a patient with ISS.

It is important to differentiate insomnia from sleep deprivation and insufficient sleep. Patients with insomnia do not generally report sleepiness when sedentary. If EDS is present in the setting of insomnia, a suspicion for an underlying sleep disorder such as OSA should be considered.

6.1.2.3.2 Obstructive Sleep Apnea

Although OSA has been discussed in the previous section, it is one of the most common causes of EDS. Most patients with OSA-associated EDS improve with treatment.

6.1.2.3.3 Narcolepsy

Narcolepsy is a relatively rare autoimmune disease, with an estimated prevalence of about 19–56 per 100,000 people in Europe and the United States.[26] Narcolepsy is characterized by the following clinical features:

- Irresistible attacks of sleep are present in all patients. Other features of narcolepsy may not be present in all patients.
- Cataplexy is characterized by sudden bilateral loss of muscle tone brought on by emotions, which can be limited to certain muscles or generalized. Although these attacks may cause the patient to fall down, full consciousness is maintained. Cataplexy is pathognomonic for narcolepsy, though it is not present in all patients. If cataplexy is present, the patient is diagnosed with narcolepsy type 1. A diagnosis of narcolepsy without cataplexy (narcolepsy type 2) is appropriate when EDS is present with REM phenomenology (hypnagogic hallucinations and sleep paralysis) but without cataplexy.
- Hypnagogic hallucination: Vivid dreams that occur at the transition from wakefulness to sleep (hypnagogic) or from sleep to wakefulness (hypnopompic).
- Sleep paralysis: A temporary inability to move or speak upon waking up or falling asleep.
- Interrupted or fragmented sleep: Patients with narcolepsy may also report experiencing fragmented sleep.

The diagnosis of narcolepsy is confirmed via overnight sleep studies (PSG), followed by multiple sleep latency testing (MSLT). The MSLT begins 1.5–2 hours after waking up in the morning. The patient is given four to five chances to nap, each separated by 2 hours. If the patient falls asleep, he is allowed to sleep for 15 minutes. Sleep latency and sleep onset REM (SOREM) are monitored. The presence of a short sleep latency (<8 minutes) and two or more SOREM support the diagnosis of narcolepsy.

OSA, periodic leg movement during sleep, and restless legs syndrome (RLS) commonly coexist in patients with narcolepsy.[26–28]

The management of patients with narcolepsy and cataplexy aims to improve daytime sleepiness and control cataplexy. Behavioral therapy and medication are used to treat the irresistible attacks of sleep. Good sleep hygiene, obtaining enough sleep at night, and strategic naps of a few minutes (when circumstances allow) are used. These short naps increase alertness in patients with narcolepsy for 1–2 hours. Stimulants may also be used to reduce sleepiness. Although the first-line treatment is modafinil, methylphenidate may be used in some patients who do not respond to modafinil.

First-line treatment for cataplexy involves the use of the selective serotonin reuptake inhibitor (SSRI) fluoxetine or the serotonin-norepinephrine reuptake inhibitor venlafaxine. Sodium oxybate (Xyrem) can also be used, particularly, in difficult cases.[29] Sodium oxybate is the only drug approved by the FDA to treat cataplexy.

6.1.2.3.4 Idiopathic Hypersomnia

Idiopathic hypersomnia is a rare disorder of unknown etiology. In contrast to patients with narcolepsy, patients with idiopathic hypersomnia exhibit uninterrupted sleep of long duration ($\geq$11 hour/day) without cataplexy. Short naps are not refreshing for patients with idiopathic hypersomnia. Short sleep latency ($<$8 minutes) and SOREM $\leq$1 are observed on MSLT. Idiopathic hypersomnia is often a diagnosis of exclusion made after all other causes of hypersomnia have been ruled out.

6.1.2.3.5 Circadian Rhythm Disorders

Circadian rhythm disorders can present with EDS (or insomnia) and may receive inaccurate diagnoses. Thus, a detailed sleep history and sleep diary/actigraphy are essential for reaching the correct diagnosis. Sleep problems arise in this group of patients due to misalignment between the body's endogenous circadian clock and the schedules imposed by society. Delayed sleep phase disorder is not uncommon among young adults between the ages of 15 and 25 years. Patients with delayed sleep phase disorder report an inability to fall asleep at night, daytime sleepiness, and difficulty waking up in the morning. Fig. 6.3B shows the sleep diary of a patient with delayed sleep phase disorder.

Shift work disorder is another common circadian rhythm disorder. Night shifts are usually the most disruptive to circadian rhythms, though early morning shifts can also be disruptive. If a circadian rhythm disorder is suspected, it is recommended that patients be referred to a sleep specialist for appropriate management of the condition, which involves behavioral therapy, light therapy, and melatonin supplementation.

6.1.2.3.6 Medical Disorders Associated With Excessive Daytime Sleepiness

Several medical disorders are either directly or indirectly associated with EDS. Therefore, medical history questions should thoroughly screen for preexisting conditions that may produce EDS. Thyroid dysfunction or diabetes can cause EDS, and previous research has suggested that OSA is more common among patients with hypothyroidism and diabetes.[30,31] Several other medical disorders have been linked to an increased risk for OSA, such as hypertension, ischemic heart disease, heart failure, atrial fibrillation, and pulmonary hypertension.[32] Patients with congestive heart failure and low ejection fraction are also at increased risk for central sleep apnea. Sleep apnea and stroke have a

TABLE 6.3 Medication Classes Commonly Associated With Daytime Sleepiness

- Alpha-adrenergic blocking agents
- Antitussives
- Anticonvulsants
- Antidepressants (monoamine oxidase tricyclic agents, other anxiolytics, selective serotonin reuptake inhibitors)
- Antipsychotics
- Beta-adrenergic blocking agents
- Antidiarrheal agents
- Genitourinary smooth muscle
- Antiemetics
- Antihistamines
- Opiate agonists and partial agonists
- Antimuscarinics and antispasmodics
- Antiparkinsonian agents
- Skeletal muscle relaxants

bidirectional relationship, such that sleep apnea increases the risk of stroke, and stroke increases the risk of developing obstructive or central sleep apnea.[33] Moreover, patients with neurodegenerative diseases such as Parkinson's disease and Alzheimer's disease usually report experiencing EDS.[34] Furthermore, several prescription medications can either impair nighttime sleep or cause daytime sleepiness (Table 6.3).

6.1.2.3.7 Psychiatric Disorders Associated With Excessive Daytime Sleepiness

Some psychiatric disorders such as depression or bipolar disorder may present with EDS. EDS can be due to a psychiatric condition, psychiatric medications, or a comorbid sleep disorder. Patients with severe depression may also present with prolonged sleep or EDS,[35] and antidepressant and antipsychotic medications can lead to daytime sleepiness.[36]

6.1.3 Restless Legs Syndrome

RLS is a sensory-motor disorder characterized by unpleasant "creepy-crawly" sensations in the lower limbs. Movement of the legs temporarily relieves these symptoms but disrupts the ability to stay asleep during nighttime, resulting in delayed or fragmented sleep.[37] The major symptoms of RLS are very disturbing sensations in the limbs (98%), and sleep disturbances are often the primary complaint (95%). Patients may report experiencing crawling, aching, tingling, or indescribable sensations in their legs, in addition to restlessness and an urge to move.

The International Restless Legs Syndrome Study Group (IRLSSG) suggested four diagnostic criteria for RLS: (1) an urge to move the legs, usually accompanied or caused by an uncomfortable sensation in the legs; (2) beginning or worsening of symptoms during periods of rest or inactivity; (3) partial or total relief of symptoms by movement; and (4) symptoms that are worse in the evening or night compared with during the day or that occur only in the evening or night (following a circadian rhythm).[38] Roughly 60% of patients with RLS are estimated to have a positive family history; furthermore, genetic association studies have linked 5 genes and 10 different alleles to RLS.[39] Although RLS is a distinct disorder, physicians often ascribe its symptoms to muscle cramps, fatigue, stress, neurological disorders, or psychiatric disorders such as depression and anxiety, which in turn result in undertreatment or inappropriate treatment of the disorder.[40,41] The 2001 Sleep in America Poll revealed that only 3% of participants with RLS symptoms had been diagnosed by a physician.[42]

The symptoms of RLS follow the circadian fluctuation of dopamine in the substantia nigra and the putamen. RLS patients have lower dopamine and iron levels in the substantia nigra and therefore respond to both dopaminergic therapy and iron administration.[43]

An RLS prevalence of 3.2%–12% has been reported in different countries.[41]

Patients with primary (idiopathic) RLS often have a genetic predisposition to the condition, which is more common in young people and usually more difficult to treat than secondary RLS. Secondary RLS has been linked to several comorbid conditions including iron deficiency, renal failure (uremia), diabetes mellitus, neurological disorders such as multiple sclerosis and Parkinson's disease, and rheumatologic diseases such as rheumatoid arthritis.[39] Moreover, the use of tricyclic antidepressants (except bupropion), SSRIs, lithium, antihistamines, and dopamine-antagonists has been associated with RLS.[44] In some patients, symptoms of RLS may worsen with nicotine, alcohol, or caffeine. It must be noted that with the advancement of understanding of pathophysiology, distinction between primary and secondary RLS has become less conspicuous. Rather, these disorders are now considered as a "precipitant" of RLS through epigenetic mechanisms.

6.1.3.1 Diagnosis

There are no specific diagnostic tests for RLS. RLS is diagnosed based on clinical findings, and PSG is not required to diagnose RLS. If iron or vitamin D deficiency is suspected, serum ferritin and vitamin D levels should be measured.

6.1.3.2 Treatment

The goal of RLS treatment is to ensure uninterrupted sleep with minimal sleep latency. If a patient is on medications that are known to be

and multiple systems atrophy and usually occurs in older men. It has also been reported in patients with narcolepsy, and in association with alcohol and some medications such as antidepressants and β-blockers.[53] During RBD, there is an absence of the normal atonia that accompanies REM sleep; therefore, patients act their dreams. Symptoms may rarely include a violent dream enactment behavior owing to a loss of atonia during REM sleep. However, in most cases, there could be minimal movements during REM sleep.

6.1.5.3 Treatment

Treatment for NREM and REM sleep parasomnias often includes avoidance of potential triggers. Additionally, care should be taken in patients with somnambulism and RBD to avoid patients' or bed-partners' injury. This necessitates a safe, well-monitored sleeping environment. There are no treatment guidelines based on randomized controlled trials for treating parasomnias. However, based on anecdotes and case series, the most common pharmaceutical treatment for REM and NREM sleep parasomnias is usually a longer acting benzodiazepine.

Small case series and case reports described the efficacy of a wide range of medications in REM parasomnias, including clonazepam, melatonin, pramipexole, acetylcholinesterase inhibitors, paroxetine, L-DOPA, zopiclone, temazepam, triazolam, alprazolam, Yi-Gan San, desipramine, carbamazepine, clozapine, and sodium oxybate.[53]

References

1. AASM. *International classification of sleep disorders (ICSD-3)*. 3rd ed. Chicago, IL: American Academy of Sleep Medicine; 2014.
2. Punjabi NM. The epidemiology of adult obstructive sleep apnea. *Proc Am Thorac Soc* 2008;**5**(2):136–43.
3. Young T, Palta M, Dempsey J, Skatrud J, Weber S, Badr S. The occurrence of sleep-disordered breathing among middle-aged adults. *N Engl J Med* 1993;**328**(17):1230–5.
4. Kapur V, Strohl KP, Redline S, Iber C, O'Connor G, Nieto J. Underdiagnosis of sleep apnea syndrome in U.S. communities. *Sleep Breath* 2002;**6**(2):49–54.
5. Alotair H, BaHammam A. Gender differences in Saudi patients with obstructive sleep apnea. *Sleep Breath* 2008;**12**:323–9.
6. BaHammam AS, Alrajeh MS, Al-Ibrahim FS, Arafah MA, Sharif MM. Prevalence of symptoms and risk of sleep apnea in middle-aged Saudi women in primary care. *Saudi Med J* 2009;**30**(12):1572–6.
7. Rahaghi F, Basner RC. Delayed diagnosis of obstructive sleep apnea: don't ask, don't tell. *Sleep Breath* 1999;**3**(4):119–24.
8. Lavie P. Sleep medicine—time for a change. *J Clin Sleep Med* 2006;**2**(2):207–11.
9. Al-Jawder SE, BaHammam AS. Comorbid insomnia in sleep-related breathing disorders: an under-recognized association. *Sleep Breath* 2012;**16**(2):295–304.
10. BaHammam AS, Al-Shimemeri SA, Salama RI, Sharif MM. Clinical and polysomnographic characteristics and response to continuous positive airway pressure therapy in obstructive sleep apnea patients with nightmares. *Sleep Med* 2013;**14**(2):149–54.

11. BaHammam AS, Al-Aqeel AM, Alhedyani AA, Al-Obaid GI, Al-Owais MM, Olaish AH. The validity and reliability of an Arabic version of the STOP-Bang Questionnaire for identifying obstructive sleep apnea. *Open Respir Med J* 2015;**9**:22–9.
12. BaHammam AS, Al-Rajeh MS, Al-Ibrahim FS, Arafah MA, Sharif MM. Prevalence of symptoms and risk of sleep apnea in middle-aged Saudi women in primary care. *Saudi Med J* 2009;**30**(12):1572–6.
13. Aldabal L, BaHammam AS. Metabolic, endocrine, and immune consequences of sleep deprivation. *Open Respir Med J* 2011;**5**:31–43.
14. Kushida CA, Littner MR, Morgenthaler T, Alessi CA, Bailey D, Coleman Jr J, et al. Practice parameters for the indications for polysomnography and related procedures: an update for 2005. *Sleep* 2005;**28**(4):499–521.
15. Epstein LJ, Kristo D, Strollo Jr. PJ, Friedman N, Malhotra A, Patil SP, et al. Clinical guideline for the evaluation, management and long-term care of obstructive sleep apnea in adults. *J Clin Sleep Med* 2009;**5**(3):263–76.
16. BaHammam AS, Alassiri SS, Al-Adab AH, Alsadhan IM, Altheyab AM, Alrayes AH, et al. Long-term compliance with continuous positive airway pressure in Saudi patients with obstructive sleep apnea. A prospective cohort study. *Saudi Med J* 2015;**36**(8):911–19.
17. Randerath WJ, Verbraecken J, Andreas S, Bettega G, Boudewyns A, Hamans E, et al. Non-CPAP therapies in obstructive sleep apnoea. *Eur Respir J* 2011;**37**(5):1000–28.
18. Ohayon MM. From wakefulness to excessive sleepiness: what we know and still need to know. *Sleep Med Rev* 2008;**12**(2):129–41.
19. Neu D, Linkowski P, le Bon O. Clinical complaints of daytime sleepiness and fatigue: how to distinguish and treat them, especially when they become 'excessive' or 'chronic'? *Acta Neurol Belg* 2010;**110**(1):15–25.
20. Chaudhuri KR. Nocturnal symptom complex in PD and its management. *Neurology* 2003;**61**(6 Suppl. 3):S17–23.
21. Gupta R, Pandi-Perumal SR, Almeneessier AS, BaHammam AS. Hypersomnolence and traffic safety. *Sleep Med Clin* 2017;**12**(3):489–99.
22. Johns MW. A new method for measuring daytime sleepiness: the Epworth sleepiness scale. *Sleep* 1991;**14**(6):540–5.
23. Adamkova V, Hubacek JA, Lanska V, Vrablik M, Kralova Lesna I, Suchanek P, et al. Association between duration of the sleep and body weight. *Physiol Res* 2009;**58**(Suppl. 1):S27–31.
24. Van Dongen HP, Maislin G, Mullington JM, Dinges DF. The cumulative cost of additional wakefulness: dose-response effects on neurobehavioral functions and sleep physiology from chronic sleep restriction and total sleep deprivation. *Sleep* 2003;**26**(2):117–26.
25. Liu Y, Wheaton AG, Chapman DP, Cunningham TJ, Lu H, Croft JB. Prevalence of healthy sleep duration among adults--United States, 2014. *MMWR Morb Mortal Wkly Rep* 2016;**65**(6):137–41.
26. Plazzi G, Ferri R, Antelmi E, Bayard S, Franceschini C, Cosentino FI, et al. Restless legs syndrome is frequent in narcolepsy with cataplexy patients. *Sleep* 2010;**33**(5):689–94.
27. BaHammam AS, Alenezi AM. Narcolepsy in Saudi Arabia. Demographic and clinical perspective of an under-recognized disorder. *Saudi Med J* 2006;**27**(9):1352–7.
28. BaHammam A. Periodic leg movements in narcolepsy patients: impact on sleep architecture. *Acta Neurol Scand* 2007;**115**(5):351–5.
29. BaHammam AS, Pandi-Perumal SR, Neubauer DN. Sodium oxybate (Xyrem®): a new and effective treatment for narcolepsy with cataplexy. In: Guglietta A, editor. *Drug treatment of sleep disorders: milestones in drug therapy*. Switzerland: Springer International Publishing; 2015. p. 231–48.

30. Misiolek M, Marek B, Namyslowski G, Scierski W, Zwirska-Korczala K, Kazmierczak-Zagorska Z, et al. Sleep apnea syndrome and snoring in patients with hypothyroidism with relation to overweight. *J Physiol Pharmacol* 2007;**58**(Suppl. 1):77–85.
31. Chasens ER, Umlauf MG, Weaver TE. Sleepiness, physical activity, and functional outcomes in veterans with type 2 diabetes. *Appl Nurs Res* 2009;**22**(3):176–82.
32. Vrints H, Shivalkar B, Hilde H, Vanderveken OM, Hamans E, Van de Heyning P, et al. Cardiovascular mechanisms and consequences of obstructive sleep apnoea. *Acta Clin Belg* 2013;**68**(3):169–78.
33. Redline S, Yenokyan G, Gottlieb DJ, Shahar E, O'Connor GT, Resnick HE, et al. Obstructive sleep apnea-hypopnea and incident stroke: the sleep heart health study. *Am J Respir Crit Care Med* 2010;**182**(2):269–77.
34. Marinus J, Visser M, van Hilten JJ, Lammers GJ, Stiggelbout AM. Assessment of sleep and sleepiness in Parkinson disease. *Sleep* 2003;**26**(8):1049–54.
35. Posternak MA, Zimmerman M. Symptoms of atypical depression. *Psychiatry Res* 2001;**104**(2):175–81.
36. Wichniak A, Wierzbicka A, Jernajczyk W. Sleep and antidepressant treatment. *Curr Pharm Des* 2012;**18**(36):5802–17.
37. Walters AS. Toward a better definition of the restless legs syndrome. The International Restless Legs Syndrome Study Group. *Mov Disord* 1995;**10**(5):634–42.
38. Allen RP, Picchietti D, Hening WA, Trenkwalder C, Walters AS, Montplaisi J. Restless legs syndrome: diagnostic criteria, special considerations, and epidemiology. A report from the restless legs syndrome diagnosis and epidemiology workshop at the National Institutes of Health. *Sleep Med* 2003;**4**(2):101–19.
39. Salas RE, Gamaldo CE, Allen RP. Update in restless legs syndrome. *Curr Opin Neurol* 2010. Available from: https://doi.org/10.1097/WCO.0b013e32833bcdd8.
40. Garcia-Borreguero D, Egatz R, Winkelmann J, Berger K. Epidemiology of restless legs syndrome: the current status. *Sleep Med Rev* 2006;**10**(3):153–67.
41. BaHammam A, Al-shahrani K, Al-zahrani S, Al-shammari A, Al-amri N, Sharif M. The prevalence of restless legs syndrome in adult Saudis attending primary health care. *Gen Hosp Psychiatry* 2011;**33**(2):102–6.
42. National Sleep Foundation. *The 2001 sleep in America Poll*. Washington, DC: National Sleep Foundation; 2001.
43. Facheris MF, Hicks AA, Pramstaller PP, Pichler I. Update on the management of restless legs syndrome: existing and emerging treatment options. *Nat Sci Sleep* 2010;**2**:199–212.
44. Rottach KG, Schaner BM, Kirch MH, Zivotofsky AZ, Teufel LM, Gallwitz T, et al. Restless legs syndrome as side effect of second generation antidepressants. *J Psychiatr Res* 2008;**43**(1):70–5.
45. Summers MO, Crisostomo MI, Stepanski EJ. Recent developments in the classification, evaluation, and treatment of insomnia. *Chest* 2006;**130**(1):276–86.
46. Almeneessier AS, Alamri BN, Alzahrani FR, Sharif MM, Pandi-Perumal SR, BaHammam AS. Insomnia in primary care settings: still overlooked and undertreated? *J Nat Sci Med* 2018;**1**(2):64–8.
47. Riemann D, Nissen C, Palagini L, Otte A, Perlis ML, Spiegelhalder K. The neurobiology, investigation, and treatment of chronic insomnia. *Lancet Neurol* 2015;**14** (5):547–58.
48. Smith MT, Perlis ML, Park A, Smith MS, Pennington J, Giles DE, et al. Comparative meta-analysis of pharmacotherapy and behavior therapy for persistent insomnia. *Am J Psychiatry* 2002;**159**(1):5–11.
49. Neubauer DN, Pandi-Perumal SR, Spence DW, Buttoo K, Monti JM. Pharmacotherapy of insomnia. *J Cent Nerv Syst Dis* 2018;**10** 1179573518770672.

50. Pagel JF, Pandi-Perumal SR, Monti JM. Treating insomnia with medications. *Sleep Sci Pract* 2018;**2**(5):1–12.
51. Pandi-perumal SR, BaHammam AS, Shapiro CM. Parasomnias. In: Price LH, Stolerman IP, editors. *Encyclopedia of psychopharmacology*. Berlin Heidelberg: Springer-Verlag; 2015. p. 1210–18.
52. Almeneessier AS, BaHammam AS. A snoring man with an abnormal sexual behavior. *J Nat Sci Med* 2018;**1**(2):85–8 Available from: https://doi.org/10.4103/JNSM.JNSM_36_18.
53. Aurora RN, Zak RS, Maganti RK, Auerbach SH, Casey KR, Chowdhuri S, et al. Best practice guide for the treatment of REM sleep behavior disorder (RBD). *J Clin Sleep Med* 2010;**6**(1):85–95.

CHAPTER

7

Obesity and Sleep Disturbances

Jacqueline Danisi, Julio Fernandez-Mendoza, Alexandros N. Vgontzas and Edward O. Bixler

Sleep Research & Treatment Center, Penn State Health Milton S. Hershey Medical Center, Penn State College of Medicine, Hershey, PA, United States

OUTLINE

7.1 INTRODUCTION: THE OBESITY EPIDEMIC

The obesity epidemic is putting an increasing strain on national healthcare systems worldwide. Obesity is defined as an excess

Molecular Sleep–Wake Cycle
DOI: https://doi.org/10.1016/B978-0-12-816430-3.00007-5

accumulation of adipose tissue that impacts the patient's health and was not considered to be a chronic disease by the Centers for Medicare and Medicaid Services (CMS) until 2004.[1,2] Unlike many other chronic diseases, obesity intervention requires changes not only in metabolism but also lifestyle as it pertains to diet and physical activity levels.

The prevalence of obesity has been increasing since the mid-1990s, when at the time, only about 15% of the adult population was categorized as obese in the United States. As of 2013, the Center for Disease Control and Prevention (CDC) estimated that more than 70% of adults were overweight, with 30% of adults being obese, and 14% categorized as severely obese.[2,3] In 2016, 46 states in the United States had obesity rates 25% or higher, of which 25 of these states had obesity rates exceeding 30% of the state adult population and 5 states had rates higher than 35%.[4] This increase in obesity prevalence across the United States has not only affected the health of individual patients, but ultimately has impacted the healthcare system as a whole. The comorbidities of obesity have resulted in mass healthcare expenditure through Medicaid, Medicare, and private insurance companies. In 2013, over $69 billion was spent on obesity-related healthcare costs, with over 43% of this cost being paid by government-supported healthcare plans, such as Medicare and Medicaid.[2] Importantly, obesity impacts the individual's quality of life and increases the risk of morbidity and mortality.

Unfortunately, childhood obesity has also shown a steady increase, comparable with that witnessed in adults. In 1963, childhood obesity was almost unheard of, with <5% of the US pediatric population suffering from the disorder, while the CDC estimated as of 2008 that more than 20% of the pediatric population was obese.[5] By 2014, more than 17% of 2–19 year olds were categorized as obese, with African-American and Hispanic youth showing increased prevalence rates of obesity at 19.5% and 21%, respectively.[6] It is projected that the lifetime medical costs of a child with obesity throughout adulthood ranges from $12,660 to $19,630 in additional expenses per child.[7] In addition to extra expenditure, childhood-onset obesity affects the patient during a critical developmental period, resulting in longitudinal health concerns into adulthood. Previous studies have shown that 77%–92% of obese adolescents remain obese in adulthood.[7] Therefore, studying the sequelae of obesity should not be limited to adult or adolescent populations, but rather, it should incorporate the metabolic effects seen in obese children.

Overweight and obese children not only have co-occurring cardiometabolic disorders, but they are also at an increased longitudinal risk of such disorders in adolescence and adulthood. In other words, cardiovascular morbidity related to obesity in adulthood may very well have its origins in childhood.[8] Combating childhood overweight and obesity is imperative to prevent cardiometabolic disorders, such as hypertension and type 2 diabetes (T2D), cardiovascular disease (CVD), or cerebrovascular disease later in life. Implementing early prevention strategies into overall wellness checkups can help lessen the burden of childhood overweight and obesity before it can prospectively impact the health of the child. With a growing population of obese patients in the United States, which more heavily affects lower income and minority populations, public health approaches should combat this widespread chronic disease from multiple perspectives. Therefore, the increase in obesity does not only involve the accumulation of adipose tissue, but it also involves the onset of adverse health outcomes. For this reason, lifestyle changes, such as diet, exercise, and sleep behavior, should be implemented into the routine of both children and adults to prevent unwanted weight gain and detrimental systemic effects.

Although overall weight gain has adverse side effects, central obesity is the most forbearing of all adiposity distributions. Central obesity is defined as the accumulation of adipose tissue in the android region, including subcutaneous adipose tissue (SAT) and visceral adipose tissue (VAT).[9] It is usually measured taking a waist circumference, via a tape measure, and compared with a waist-to-hip ratio. Unlike other forms of adipose distribution, central obesity is found not only in individuals that are categorized as obese by BMI standards, but also among overweight individuals. For example, nonobese (BMI < 30) men with central obesity (>102 cm waist circumference) have an 87% higher mortality risk than men with similar BMI but no central obesity.[10] Abdominal fat tissue, especially VAT, is more lipolytic and metabolically relevant than other fat tissues, and therefore a more prominent risk factor for insulin resistance and other cardiometabolic dysfunctions. Therefore, the development of obesity, especially central obesity, is important to understand in order to prevent and treat this weight imbalance.

7.2 ETIOLOGY AND PATHOPHYSIOLOGY OF OBESITY

To understand the adverse health outcomes of obesity and its relationship with disturbed and insufficient sleep, we must first understand why and how obesity has become such a large epidemic through unmodifiable and modifiable causes as well as how it becomes a chronic condition. The etiological and pathophysiological mechanisms of weight gain and obesity are summarized in Fig. 7.1.

Modifiable causes imply that the individual has control over these factors, while unmodifiable causes are beyond the control of the individual. Childhood-onset obesity may be caused by a multitude of

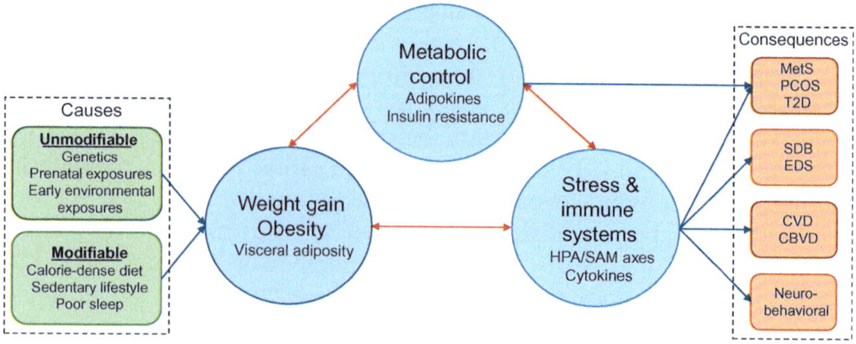

FIGURE 7.1 **Etiology, pathophysiology, and consequences of obesity.** Etiological factors contributing to weight gain and the development of obesity are depicted in *green boxes*, while pathophysiologic mechanisms are depicted in *blue circles*. The latter include adipokines dysregulation (leptin, ghrelin, and adiponectin), insulin resistance, low-grade inflammation (cytokines), and stress system activation (cortisol, norepinephrine). *Bidirectional arrows* represent the self-perpetuating vicious cycle of weight gain, metabolic changes, stress activation, and low-grade inflammation. *CBVD*, cerebrovascular diseases; *CVD*, cardiovascular disease, including hypertension; *EDS*, excessive daytime sleepiness; *HPA*, hypothalamic–pituitary–adrenal; *MetS*, metabolic syndrome; *PCOS*, polycystic ovary syndrome; *SAM*, sympathetic–adrenal–medullary; *SDB*, sleep disordered breathing; *T2D*, type 2 diabetes.

unmodifiable biological factors such as genetic heritability, prenatal and early environmental exposures, as well as early-onset endocrine disorders.[11] Environmental risk factors, such as exposure to gestational diabetes in utero or a shorter breastfeeding duration, have been shown to increase a child's risk for obesity.[12] The main cause of obesity throughout life, however, is an energy imbalance in regard to diet and physical activity. One of the primary modifiable causes feeding into the obesity epidemic is the change to a predominantly sedentary lifestyle. This is mostly due to the fact that the use of computers has increased exponentially in both the workplace and the home.[13] This decrease in physical activity generates a never-ending spiral of obesity because it leads to health consequences that further limit physical behavior, and thus increases the patient's sedentary lifestyle.[14]

This sedentary lifestyle alone is not the only cause for this increase in obesity, but it is exacerbated by a change in the Western diet. The Western diet now encompasses a diet high in simple carbohydrates, saturated fat, and sugar, all of which have been linked to an increased risk of heart disease.[15] This decrease in physical activity, combined with an increase in low-nutrient, dense caloric intake, has directly aided in the obesity epidemic via an imbalance in energy intake and energy expenditure.[16] In fact, the relative contribution of a sedentary lifestyle and increased caloric intake has been shown to affect the weight of children more dramatically than that of adults.[17] Additionally, inadequate and poor sleep may play a role in the onset, progression, or chronicity of obesity when combined with other unmodifiable and modifiable factors, as reviewed below.

7.3 INSUFFICIENT AND DISTURBED SLEEP IN OBESITY

Sleep is a critical modulator of neuroendocrine function and cardiometabolic health. As the prevalence of obesity continues to rise in both children and adults, it is crucial to identify how sleep may play a role in obesity's onset and maintenance. Additionally, some sleep disorders, such as obstructive sleep apnea (OSA), show a higher prevalence in obese individuals. Inadequate sleep has been linked to metabolic disorders such as T2D and insulin resistance, and emotional disorders that may further impact weight gain. The mechanistic relationship between obesity and sleep, however, is not exactly clear. Therefore, the role of sleep in the development of obesity must be further explored.

7.3.1 Obesity and Sleep Duration

As mentioned previously, sleep is important for maintaining physical, mental, and cognitive health. Children aged 6–12 years old should obtain 9–10 hours of sleep a night, while adolescents are advised to sleep around 8 hours a night.[18] Adults should sleep at least 7 hours a night on a regular basis to promote a healthy lifestyle.[19] In an analysis carried out by the CDC in 2009, 35.3% of US adults had less than 7 hours of sleep per night, compared with approximately 29% in the 2004–06 National Health Interview Survey.[20] Thus, early childhood intervention strategies are important to help establish a healthy lifestyle, such as adequate and satisfactory sleep.

Short sleep duration, as measured by self-reports, has been shown to be associated with the development of obesity.[21] Factors such as lifestyle behaviors, sleep complaints or stress, or even socioeconomic status influence self-reports of short sleep duration or increased habitual sleep variability (HSV), which is characterized by a fluctuation in night-to-night sleep duration across the week. Short sleep may lead to feelings of low energy, and to compensate for this lack of energy, individuals may consume more calories. This excess caloric intake over time leads to an imbalance between energy consumption and expenditure.[22] Additionally, it has been shown that sleep deprived individuals are more likely to consume diets higher in simple carbohydrates, saturated fat, and refined sugar.[23] Therefore, not only are these individuals consuming more calories, but these additional calories are from energy-dense foods. Another study found that very short sleepers (less than 5 hours) had an additional intake of 178 kcal/day compared with individuals who received at least 8 hours of sleep.[24] Although this increase in calories may be small, this equates to an additional 1200 kcal in a single week.

However, other studies have shown that increased HSV, as measured by 7 days of at-home actigraphy, was associated with increased energy consumption, particularly from fats and carbohydrates, independent of average habitual sleep duration (HSD).[25] Another study observed that in adolescents, increased HSV, but not average HSD, was associated with abdominal adiposity and that such association could be partially attributed to excessive food consumption.[26] Additionally, individuals who report short sleep duration are more likely to be "snackers" and eat fewer scheduled meals throughout the day. This snacking behavior has been observed predominately in short (5–6 hours of sleep) and very short (less than 5 hours of sleep) sleepers. These short sleepers also tended to have less variety of food in their diet leading to lack of essential vitamins and nutrients.[27] Interestingly, another study found a more consistent and stronger relationship between increased HSV and snack consumption during weekend/vacation days, when the participants sleep longer than during weekdays.[25] Therefore, inadequate sleep combined with increased caloric intake may aid in the prevalence of obesity.

Increased HSV is seen in working adults and full-time students, where they experience insufficient sleep during the workweek and catch-up sleep on the weekends.[28] During the workweek the individual may be consuming more calories to compensate for sleep debt, and decreasing physical activity on the weekends when they are experiencing longer sleep duration. This endless cycle of insufficient sleep and catch-up sleep can aid in the accumulation of adipose tissue and subsequent weight gain, however, the underlying mechanism of this relationship remains poorly understood. Nevertheless, it appears that increased HSV is most relevant to the maintenance of body weight. For example, one study showed that with each hour increase in HSV, fat content increased by more than 10% of the total fat in the corresponding region, e.g., VAT.[26] This phenomenon is also seen in individuals that work unconventional hours, such as third shifts or rotating shifts. The change in the circadian rhythm can disrupt sleep variation, and result in catch-up sleep during the daytime hours when physical activity is usually high for most individuals. Therefore, individuals who work on unconventional schedules may have a compounding effect of sleep debt, sleep variability, and circadian misalignment, that when added to decreased physical activity, may foster weight gain.[21]

Although self-reported sleep duration is associated with obesity, this phenomenon is not always found in studies using objective sleep measures (i.e., PSG or actigraphy).[29] There seems to be a discrepancy between objective sleep duration and reported subjective sleep duration. In these cases where objective and subjective sleep duration do not match, it is probable that other factors are at play causing feelings of short sleep.[30] In these situations, it is important to identify an alternative cause for subjective reports of short sleep. For example, one study showed that both obese males and females who reported poor sleep (i.e., moderate-to-severe difficulties falling or staying asleep, early morning awakening, and/or nonrestorative sleep) also reported shorter sleep duration compared with obese individuals who did not have complaints of poor sleep.[31] Therefore, self-reported short sleep duration appears to be a surrogate of poor sleep and sleep complaints.

A large, population-based study found that self-reported short sleep duration was associated with the development of obesity.[29] After controlling for poor sleep complaints and level of emotional stress, however, the significant association between self-reported short sleep duration and incident obesity disappeared.[29] In fact, these two factors of poor sleep and emotional stress were strong predictors of incident obesity and played an additive role between each other. Specifically, the individuals who had poor sleep complaints at baseline and developed obesity had the highest levels of emotional stress and reported the shortest sleep duration at baseline when they were nonobese.[29,32] Interestingly, the association of emotional stress with incident obesity was stronger in young and middle-aged adults, while poor sleep was associated with incident obesity across all age groups. Furthermore, the results stratified by gender showed that the effect of emotional stress was stronger in women, while poor sleep was a strong predictor of incident obesity in men.[29] Subjective short sleep duration may be the result of emotional stress, and not directly linked to the patient's obesity.

Therefore, understanding the nature of self-reported short sleep duration is important and, thus, disentangling sleep disturbances and level of emotional stress should be considered to better understand the relationship between insufficient sleep and obesity. Additionally, self-reported short sleep duration may actually precede the development of emotional disorders that may cause an individual to experience weight gain. As reasoned previously, emotional stress may lead to the development of poor sleep. If an individual is experiencing sleep difficulties, anxiety, or depression, they may turn to food for comfort. Metabolic imbalances worsened by obesity also impact behavioral health, and overweight or obese individuals are believed to be at a greater risk for depression.[33] Social stigmas surrounding weight, particularly in women, may heighten depression, resulting in a never-ending cycle of emotional distress.[34] Therefore, emotional distress and psychiatric disorders may play a significant role in the duration and severity of weight gain due to shortened sleep duration.

In summary, in addition to short or highly variable habitual sleep, poor sleep and associated emotional distress have shown to have similar effects on the progression to obesity and explain a high proportion of the variance in the association of self-reported short sleep duration with the development of obesity.

7.3.2 Obesity and Poor Sleep

Poor sleep, like difficulties falling and/or staying asleep or night-to-night sleep variability, is associated with increased weight gain and subsequent adverse cardiometabolic outcomes. There are many factors that can lead to poor sleep, such as stress, physiological changes, lifestyle choices, or environmental exposures. Early on, it was suggested that obesity-related poor sleep and subjective fatigue appeared to be the result of emotional stress.[35] Additionally, individuals who are obese may be at an increased risk for incident poor sleep. A large, population-based study found that obesity was an independent risk factor for developing poor sleep,[36] thus indicating that overweight individuals are at an increased risk of developing poor sleep in comparison with normal weight individuals.

Some underlying physiological mechanisms corresponding to overweight individuals, therefore, may be the precursor causing these sleep abnormalities. This suggests that disrupted sleep may not only be related to individual routine or lifestyle changes. Additionally, individuals who experience poor sleep may consume more calories during the day to compensate for their lack of energy. As discussed previously, this endless cycle will lead to increased weight gain, further poor sleep, and increased calorie intake and metabolic dysregulation. Therefore, it is imperative to address sleep disturbances early on to prevent the development of obesity. The association between poor sleep and metabolic impairments suggests an important role for sleep in the growing obesity epidemic.

7.3.3 Obesity and Chronic Insomnia

Chronic insomnia is a disorder of poor sleep occurring at least 3 nights a week, associated with significant daytime functioning impairment and with a chronicity of at least 3 months, as per current diagnostic criteria.[37] Chronic insomnia, as opposed to isolated poor sleep, is a persistent condition that stems from physiological and behavioral perpetuating factors. Chronic insomnia is a disorder of 24-hour hyperarousal, as indexed by hyperactivity of the hypothalamic–pituitary–adrenal (HPA) axis, sympathetic–adrenal–medullary (SAM) axis, and central nervous system (CNS) activation.[38] Interestingly, a large, population-based study found that nonobese individuals with chronic insomnia were not significantly associated with an increased risk of developing obesity.[29] In addition, individuals who developed chronic insomnia, regardless of whether they had been good or poor sleepers at baseline, were significantly heavier at baseline when compared with good sleepers who did not develop either poor sleep or chronic insomnia.[36,39] These findings were consistent with previous studies in large clinical samples of individuals with chronic insomnia[40,41] as well as those showing an increased resting metabolic rate in nonobese with chronic insomnia as compared with normal sleepers.[42] This increased basal metabolic rate may be a direct result of the increased hyperactivity of the sympathetic nervous system, which also contributes to the poor sleep. It is plausible that nonobese individuals with chronic insomnia are not at risk of developing obesity due to their greater energy consumption. If chronic insomniacs do develop obesity, it is independent of their insomnia diagnosis and is more likely a result of other underlying metabolic disorders or lifestyle factors (e.g., diet and physical activity).

Individuals with chronic insomnia, however, are associated with elevated levels of proinflammatory cytokines.[43] This proinflammatory state, together with the HPA and SAM axis changes mentioned above, puts chronic insomniacs, particularly those with objective short sleep duration, at an increased risk for cardiometabolic disorders such as hypertension or T2D.[44–47] In summary, the evidence so far supports that chronic insomnia with objective short sleep duration is associated with increased cardiometabolic risk independent of obesity.

7.3.4 Obesity and Sleep Apnea

Sleep disorders may also develop as a result of increased weight gain, as mentioned above in the case of poor sleep. Sleep disordered breathing (SDB) is characterized as complete (apneas) or partial (hypopneas) breathing pauses and hypoxia.[48] In overweight and obese individuals, OSA is highly prevalent and is caused by an obstruction of the upper airway of the respiratory system. In obese individuals, this is most commonly a result of excess fat and tissue compressing the pharynx, preventing breathing mechanisms to occur.[49,50] It is plausible that there is a reciprocal interaction between obesity and OSA and that metabolic abnormalities in obesity and OSA are in a continuum so that the presence of OSA in overweight individuals leads to a natural disease progression towards obesity.[29] One study showed that both men and women with higher BMI values had more severe apnea, thus indicating that obesity is a significant contributing factor to sleep apnea. Interestingly, women were found to have a higher threshold for the effects of obesity on apnea.[51]

OSA is more prevalent in men, and as mentioned above, in individuals with an increased BMI, particularly in the severely obese range (>35 kg/m^2). Men may have higher prevalence of OSA because males show higher central adipose tissue distribution, while females typically show a higher level of peripheral and gynoid adipose tissue distribution. OSA in obese middle-aged men is associated with visceral obesity, inflammatory cytokine elevations, hyperleptinemia, and hyperinsulinemia.[52] This phenomenon may be due to increased VAT, which may lead to higher levels of cytokine and adipokine secretion, and the development of OSA. Men with OSA have been found to have higher plasma concentrations of the adipose tissue-derived hormone, leptin, and of the inflammatory, fatigue-causing, and insulin resistance-producing cytokines tumor necrosis factor-α (TNF-α) and interleukin-6 (IL-6) than obese men without OSA, who had intermediate values, or lean men, who had the lowest values.[52]

Additionally, one study showed that VAT is associated with SDB as early as adolescence and is not limited to middle age, leading to an earlier onset of cardiometabolic and immune dysregulation.[53] The central distribution of adiposity in males also includes higher levels of VAT, which correlates to the higher rates of CVD and other adverse health outcomes.[54] For example, elevated CRP levels have been found to be the result of increased VAT and predict the development of OSA in adolescent males but not females.[55] Additionally, one study found that it is VAT, rather than generalized obesity, that predisposes obese individuals to the development of OSA. Individuals with OSA have been found to have significantly higher amounts of VAT compared with obese subjects without OSA, while their BMI, SAT, or total body fat were similar between the two groups.[52]

Therefore, obese men who are diagnosed with OSA may be manifesting metabolic syndrome (MetS) and may be at risk of worsening some of its components (e.g., blood pressure, insulin resistance) because of their increased levels of VAT and cytokine production. Due to this, it is imperative to treat OSA as soon as possible. Continuous positive airway pressure treatment, however, has only been shown to improve sleep duration and improve metabolic health, but does not reduce obesity like previously believed.[56–58] Therefore, weight loss cannot be directly linked to increased sleep duration. In the case of OSA, the underlying mechanisms involved in its development and maintenance extends beyond simply the current body mass of the patient.

7.3.5 Obesity and Daytime Sleepiness

An individual who exhibits OSA, poor sleep, or shortened sleep may experience excessive daytime sleepiness (EDS). A shortened period of sleep can indeed lead to EDS, and some investigators have proposed that it can also increase an individual's risk of gaining weight; however, this association may be indirect. This EDS presentation may be associated with the individual's obesity, while remaining independent from OSA, poor sleep, or short sleep. One study in particular found that severe obesity, even in the absence of sleep apnea, is associated with increased daytime sleepiness.[59] Further, it was found that individuals with larger BMIs, depression, or of younger age are more likely to report more severe EDS compared with their peers.[60] Therefore, obesity or shortened sleep duration independent of obesity, may not be the only factor driving the development of EDS. In regard to a clinical setting, it is important for practitioners to consider the multiple causes of EDS, such as depression or age, that are independent of the patient's BMI.

There is the possibility that if an individual is experiencing short sleep, and subsequently consuming more calories at night, they may have a tendency to nap more during the day. This reduces physical activity, leading to further weight gain. However, this behavioral induction of EDS and obesity is unlikely to represent the widespread association. In fact, obesity and weight gain are more likely to be predictors of EDS than vice versa. Longitudinal studies have addressed this question and found that obesity, and weight gain in nonobese individuals, were strong predictors of the development and the chronicity of EDS,[61] while EDS was not significantly associated with the development of obesity.[29] These longitudinal findings were independent of the presence of OSA, poor sleep, or self-reported sleep duration, and provided further evidence that EDS is more likely a consequence of obesity rather than a premorbid risk factor for obesity.[29,61]

From a mechanistic standpoint, accumulating data suggest that metabolic aberrations and chronic low-grade inflammation, conditions common in obesity, may play a role in the pathogenesis of obesity-related EDS.[35,62] Therefore, compared with subjective EDS, objective EDS is a stronger predictor of morbidity and mortality due to its relationship with low-grade inflammation. These findings relating obesity and MetS to the development of EDS are clinically important. These metabolic imbalances should be considered by a health professional if evidence of EDS presents in a patient, especially if a patient is not experiencing poor sleep or OSA. Therefore, to alleviate EDS, an emphasis should be placed on the individual losing weight through the alteration of modifiable factors.

7.4 THE ROLE OF INFLAMMATION AND THE STRESS SYSTEM IN THE ASSOCIATION BETWEEN OBESITY AND SLEEP DISTURBANCES

As mentioned previously, inadequate sleep and obesity are both associated with chronic low-grade inflammation. Chronic low-grade inflammation is defined as a two- to threefold increase in cytokine concentration in the blood plasma that may lead to adverse health outcomes.[63] To understand why obese individuals have increased levels of inflammation, the endocrine functionality of adipose tissue must first be considered. Adipose tissue is recognized as an active endocrine organ and the main energy storage unit for the body.[64] Excess adiposity, like that seen in overweight and obese individuals, may lead to dysregulation of adipokines, adipose secretory factors that contribute to the development of low-grade systemic inflammation.[65]

Levels of cytokine secretion, however, are not equal across all types of adipose tissue distributions. Particularly, VAT is the type of adipose tissue involved in increasing inflammation levels. The unique position of VAT around the abdominal organs increases secretion of fatty acids to the liver.[64] This excess fatty acid buildup in the liver leads to higher triglyceride production, and thus an increased risk of cardiac-related diseases. Additionally, VAT's position around endocrine organs located in the abdomen, such as the pancreas, may lead to insulin resistance and other related metabolic abnormalities.[66] Thus, central obesity, independent from general obesity, is associated with a higher MetS burden.[26]

It is, however, not the adipose tissue that releases these inflammatory biomarkers, but macrophages located within the tissue. Macrophages in VAT secrete high levels of cytokines, leading to immune dysregulation. It is this immune dysregulation that has been directly linked to cardiometabolic abnormalities such as impaired glucose uptake and development of cardiovascular comorbidities.[67] Therefore, it is not the presence of the adipose tissue per se that is causing metabolic abnormalities, but rather, the excess secretion of cytokines and adipokines. As mentioned previously, OSA patients are documented to have increased inflammatory biomarker levels in the plasma, partly due to their large amounts of VAT.

There are five main inflammatory cytokines and adipokines produced and released by adipose tissue: IL-6, TNF-α, adiponectin, leptin, and ghrelin. IL-6 is a proinflammatory cytokine secreted by macrophages to stimulate an immune response in nearby tissue. When produced by adipocytes, IL-6 is thought to cause increased secretion of C-reactive protein (CRP) secreted by the liver in obese individuals.[66]

CRP is usually produced in response to high levels of inflammation, and aids in innate immunity. This system, however, can be flawed in that excess CRP will be produced in response to increasing levels of IL-6, even if there is no invading pathogen present.[68] TNF-α is also a proinflammatory cytokine that works alongside IL-6 in an immune response. However, overexpression in obese individuals may lead to insulin resistance.[69] Therefore, IL-6, CRP, and TNF-α may serve as proinflammatory biomarkers due to their mechanistic release during periods of elevated systemic inflammation.

Adiponectin, an adipokine, is a protein involved in glucose and lipid metabolism. It decreases hepatic gluconeogenesis and increases fatty acid oxidation.[70] Adiponectin has a protective effect by increasing insulin sensitivity; thus, low levels of adiponectin are found in obese individuals who exhibit symptoms of insulin resistance.[71] In fact, the proportion of adiponectin to adipose tissue is actually an inverse relationship. This means that individuals with excess adipose tissue, such as those who are overweight or obese, would have low levels of adiponectin, leading to a lack of prevention of low-grade inflammation. Abnormal levels of adiponectin, along with other cytokines, are seen in OSA patients. One study showed that adiponectin had a reciprocal relationship to the severity of OSA; as OSA severity increased, adiponectin levels decreased.[72,73]

Increased levels of these proinflammatory biomarkers are seen in OSA patients. While OSA is an independent predictor of elevated proinflammatory cytokines, there is evidence that central obesity precedes systemic inflammation, metabolic dysfunction, and subsequent development of OSA.[50] As the lipid droplets increase in size, due to weight gain, the adipocytes start to release more cytokines and adipokines. Thus, as an individual gains more weight, and subsequently develops more severe OSA, they will continue to elevate their plasma levels of IL-6, TNF-α, and CRP. One study showed that although healthy women tend to have significantly higher values of TNFR1, CRP, leptin, and adiponectin than men, men with OSA appear to have a more severe inflammatory response, independent of age and BMI.[50] This in turn, means that individuals experiencing OSA may be at risk of developing obesity-related comorbidities, such as T2D, hypertension, and dyslipidemia independent of their body mass. Therefore, OSA in the obese may also contribute to further inflammation and metabolic dysregulation.

Sleep deprivation has also been linked to higher levels of IL-6 production and imbalance of ghrelin and leptin hormones.[21,35,50,74] The hormones ghrelin and leptin regulate body weight, energy balance and hunger. Ghrelin is a hormone in charge of appetite control and energy balance.[75] This hormone signals to the hypothalamus the state of fat storage, inhibiting food intake and further adipose accumulation.[76]

Therefore, ghrelin is known as the "hunger" hormone and stimulates appetite when energy storage is low. Although in theory obese individuals should have low plasma levels of ghrelin, some studies show they are sensitive to the hormone, meaning they will feel hungry even when there is ample energy storage.[77] Not only obesity can affect ghrelin levels, but also sleep deprivation has been shown to cause similar physiological results. When participants experienced sleep deprivation, they experienced significantly higher ghrelin levels.[78] These elevated ghrelin levels can lead to feelings of hunger, causing the individual to consume more calories than necessary. Therefore, increased ghrelin levels may be the link between sleep deprivation and the development of obesity.

Another hormone, leptin, stimulates the secretion of thyroid stimulating hormone by the anterior pituitary gland, directly acting on the thyroid gland, the main metabolism control center.[79] Several studies have shown that short sleep duration reduced leptin levels and increased ghrelin levels within the plasma.[80,81] These hormonal changes after shortened sleep duration may be why individuals who experience consistent short sleep experience increases in BMI. Another study showed that there was a significant increase in daytime leptin levels after only one night of short sleep; however, there was no change in adiponectin levels.[82] This observed elevation of daytime leptin levels is consistent with observed lower HPA axis activity, which has an inhibiting effect on leptin secretion.[83] Therefore, it is possible that individuals who experience only shortened sleep, and not stressful sleep, will not have a dysregulation in hunger-related hormones.

In relation to OSA, one study suggested that the increase in leptin levels in OSA is possibly related to the higher amount of VAT and/or cytokines in this group. It was also demonstrated that not only the amount but also the daily patterns of TNF-α, IL-6, and leptin secretion were disturbed in individuals with OSA compared with obese individuals without OSA.[52] Although these inflammatory cytokines aid in local inflammation, they also play a large role in CNS regulation. For example, IL-6 and TNF-α have been shown to stimulate the HPA axis, the predominate pathway for regulating chronic stress.[84] Specifically, IL-6 is thought to be a strong instigator for the release of corticotropin-releasing factor (CRF) from the hypothalamus. CRF targets the release of adrenocorticotropic hormone from the anterior pituitary,[84] which then acts on the adrenal gland to stimulate the release of glucocorticoids, specifically cortisol—a major factor in the stress response. An elevated production of cortisol in response to physiological stress has been shown to increase the presence of VAT, hypertension, insulin resistance, and dyslipidemia.[52,85]

Additionally, the SAM axis also manages physiological stress through acute stress response. Like that seen in the HPA axis, the hypothalamus stimulates the adrenal gland; however, epinephrine and norepinephrine are secreted as a result of sympathetic stimulation.[86] As previously mentioned, chronic insomnia is a disorder of hyperarousal due to hyperactivity of the HPA and SAM axes.[38] Unlike that seen in other chronic stress conditions, chronic insomniacs are not at risk for obesity. Males diagnosed with OSA, however, are not protected from obesity. They tend to have significantly higher amounts of intraabdominal visceral fat in addition to higher levels of adipokine and cytokine production leading to an increase in overall systemic inflammation.[87]

In regard to short sleep duration, one study showed that individuals who experience short sleep with preexisting MetS may be linked to greater central autonomic and metabolic dysfunction.[88] Interestingly, sleep deprivation does not always show consistent effects on cortisol levels across studies unlike that seen in OSA. Multiple studies have shown that there is a significant increase in plasma IL-6 following both mild (6 hours of sleep per night for one workweek)[74] and moderate-to-severe (4 hours of sleep per night for 12 days) curtailment of sleep.[89] Other studies have shown increased monocyte production of IL-6[90] and levels of IL-6 mRNA[91] in the morning following sleep restriction. Based on these data, there should be a resulting increase in cortisol because of the increased IL-6 production;[21] however, this has not always been the case.[74,87,92]

In summary, the relationship between obesity and sleep disorders should integrate behavioral, neuroendocrine, and immune mechanisms. Systemic inflammation appears to be a final common pathway in the relationship between obesity and inadequate sleep, OSA, EDS, and cardiometabolic sequela. Furthermore, increased inflammation and other neuroendocrine dysregulations may lead to a compounding effect causing adverse cardiometabolic outcomes. Therefore, addressing the relationship between inadequate sleep and obesity requires strong conceptual models and causal mechanisms that can help explain their relative contribution to the global health of an individual.

7.5 CONCLUSION

The global obesity epidemic is rising due to unmodifiable and modifiable causes, one of them being inadequate sleep. Although shortened sleep duration or disturbed sleep have been shown to lead to a greater incidence of obesity, nonobese chronic insomniacs do not follow this trend. This reveals that not all types of inadequate sleep lead to weight gain and onset of subsequent obesity. In addition, OSA has shown to have a reciprocal relationship with obesity, meaning that these individuals are already on a path to develop obesity independent of the OSA.

OSA instead may aid in the progression of obesity due to the increase in the accumulation of higher plasma proinflammatory load. This increase in proinflammatory cytokines aids in the development of adverse cardiometabolic and immune dysregulation. In turn, this metabolic dysregulation may lead to the development of EDS. Thus, the weight gain may be the cause for the metabolic imbalances leading to EDS, rather than EDS being the cause of the weight gain.

References

1. Pozza C, Isidori, A. What's behind the obesity epidemic. In: Laghi A, Rengo M, editors. Springer; *Imaging in bariatric surgery*; 2018.
2. Wang Y, Pamplin J, Long M, Ward Z, et al. Severe obesity in adults cost state Medicaid programs nearly $8 billion in 2013. *Health Aff* 2015;**34**(11):1923–31.
3. Center for Disease Control and Prevention. *National Center for Health Statistics.* Retrieved April 2, 2018, from: <https://www.cdc.gov/nchs/fastats/obesity-overweight.htm>; 2017.
4. Center for Disease Control and Prevention. *Overweight & obesity*. Retrieved March 30, 2018, from: <https://www.cdc.gov/obesity/data/prevalence-maps.html>; 2017.
5. Center for Disease Control and Prevention. *CDC grand rounds: childhood obesity in the United States*. Retrieved March 30, 2018, from: <https://www.cdc.gov/mmwr/preview/mmwrhtml/mm6002a2.htm>; 2011.
6. Ogden C, Carroll M, Fryar C, Flegal K. Key findings what was the prevalence of obesity among adults in 2011–2014? *Nat Center Health Stat* 2015;**219**:1–8.
7. Finkelstein E, Graham W, Malhotra R. Lifetime direct medical costs of childhood obesity. *Pediatrics* 2014;**133**(5):854–62.
8. Reilly J, Methven E, McDowell Z, et al. Health consequences of obesity. *Arch Dis Child* 2014;**88**:748–52.
9. He F, Rodriguez-Colon S, Fernandez-Mendoza J, Vgontzas A, et al. Abdominal obesity and metabolic syndrome burden in adolescents—Penn State Children Cohort Study. *J Clin Densitom* 2015;**18**(1):30–6.
10. Sahakyan K, Somers V, Rodriguez-Escudero J, Hodge D, et al. Normal-weight central obesity: implications for total and cardiovascular mortality. *Ann Intern Med* 2015;**163** (11):827–37.
11. Gurnani M, Birken C, Hamiliton J. Childhood obesity: causes, consequences, and management. *Pediatr Clin North Am* 2015;**62**(4):821–40.
12. Bass R, Eneli I. Severe childhood obesity: an under-recognised and growing health problem. *Postgrad Med J* 2015;**91**(1081):639–45.
13. Tremblay M, Colley R, Saunders T, Healy G, et al. Physiological and health implications of a sedentary lifestyle. *Appl Physiol Nutr Metab* 2010;**35**(6):725–40.
14. Williams E, Mesidor M, Winters K, Dubbert P, et al. Overweight and obesity: prevalence, consequences, and causes of a growing public health problem. *Curr Obes Rep* 2015;**4**(3):363–70.
15. Wright S, Aronne L. Causes of obesity. *Abdom Radiol* 2012;**37**(5):730–2.
16. Simon G, Korff M, Saunders M, Miglioretti D, et al. Association between obesity and psychiatric disorders in the US adult population. *Arch Gen Psychiatry* 2006;**63** (7):824–30.
17. Styne D, Arslanian S, Connor E, Farooqi I, et al. Pediatric obesity—assessment, treatment, and prevention: an Endocrine Society Clinical Practice Guideline. *J Clin Endocrinol Metab* 2007;**102**(3):709–57.

18. Bharti B, Mehta A, Malhi P. Sleep problems in children: a guide for primary care physicians. *Indian J Pediatr* 2012;**80**(6):492–8.
19. Watson N, Badr M, Belenky G, Bliwise D, et al. Recommended amount of sleep for a healthy adult: a joint consensus statement of the American Academy of Sleep Medicine and Sleep Research Society. *Sleep* 2015;**38**(6):843–4.
20. Center for Disease Control. Unhealthy sleep-related behaviors – 12 states. *Morb Mort Wkly Rep* 2009;**60**(8):223–38.
21. Van Cauter E, Tasali E. Endocrine physiology in relation to sleep and sleep disturbances. In: Kryger MH, Roth T, Dement WC, editors. *Principles and practice of sleep medicine*. 6th ed. New York: Elsevier Academic Press; 2016. p. 202–19.
22. Markwald R, Melanson E, Smith M, Higgins J, et al. Impact of insufficient sleep on total daily energy expenditure, food intake, and weight gain. *Proc Natl Acad Sci* 2013;**110**(14):5695–700.
23. Dashti H, Scheer F, Jacques P, Lamon-Fava S, et al. Short sleep duration and dietary intake: epidemiologic evidence, mechanisms, and health implications. *Adv Nutr* 2015;**6**(6):648–59.
24. Patterson R, Emond J, Natarajan L, et al. Short sleep duration is associated with higher energy intake and expenditure among African-American and Non-Hispanic White adults. *J Nutr* 2014;**144**(4):461–6.
25. He F, Bixler E, Berg A, Kawasawa Y, et al. Habitual sleep variability, not sleep duration, is associated with caloric intake in adolescents. *Sleep Med* 2015;**16**(7):856–61.
26. He F, Bixler E, Liao J, Berg A, et al. Habitual sleep variability, mediated by nutrition intake, is associated with abnormal obesity in adolescents. *Sleep Med* 2015;**16**(120):1489–94.
27. Grandner M, Jackson N, Gerstner J, Knutson K. Dietary nutrients associated with short and long sleep duration. Data from a nationally representative sample. *Appetite* 2013;**64**:71–80.
28. Patel S, Hayes A, Blackwell T, Evans D, et al. The association between sleep patterns and obesity in older adults. *Int J Obes* 2014;**38**:1159–64.
29. Vgontzas A, Fernandez-Mendoza J, Miksiewicz T, Kritikou I, et al. Unveiling the longitudinal association between short sleep duration and the incidence of obesity: The Penn State Cohort. *Int J Obes* 2014;**18**(4):291–2.
30. Bliwise D, Young T. The parable of parabola: what the U-shaped curve can and cannot tell us about sleep. *Sleep* 2007;**30**(12):1614–15.
31. Vgontzas A, Lin H, Papaliaga M, Calhoun S, et al. Short sleep duration and obesity: the role of emotional stress and sleep disturbances. *Int J Obes* 2008;**32**:801–9.
32. Vgontzas A, Bixler E. Short sleep and obesity: are poor sleep, chronic stress, and unhealthy behaviors the link? *Sleep* 2008;**31**(9):1203.
33. LaGrotte C, Fernandez-Mendoza J, Calhoun S, Liao D, et al. The relative association of obstructive sleep apnea, obesity, and excessive daytime sleepiness with incident depression: a longitudinal, population-based study. *Int J Obes* 2016;**40**(9):1397–404 (2005).
34. Marmorstein N, Iacono W, Legrand L. Obesity and depression in adolescence and beyond: reciprocal risks. *Int J Obes* 2014;**38**(7):906–11.
35. Vgontzas A, Bixler E, Chrousos G, Pejovic S. Obesity and sleep disturbances: meaningful sub-typing of obesity. *Arch Physiol Biochem* 2008;**114**(4):224–36.
36. Fernandez-Mendoza J, Vgontzas A, Bixler E, Singareddy R, et al. Clinical and polysomnographic predictors of the natural history of poor sleep in the general population. *Sleep* 2012;**35**(5):689–97.
37. Morin C, Benca R. Chronic insomnia. *Lancet* 2012;**379**(9821):1129–41.
38. Vgontzas A, Bixler E, Lin H, Prolo P, et al. Chronic insomnia is associated with nyctohemeral activation of the hypothalamic-pituitary-adrenal axis: clinical implications. *J Clin Endocrinol Metab* 2001;**86**(8):3787–94.

39. Singareddy R, Vgontzas A, Fernandez-Mendoza J, Liao D, et al. Risk factors for incident chronic insomnia: a general population prospective study. *Sleep Med* 2012;**13**(4):346–53.
40. Huang L, Zhou J, Sun Y, et al. Polysomnographically determined sleep and body mass index in patients with insomnia. *Psychiatry Res* 2013;**209**(3):540–4.
41. Crönlein T, Langguth B, Busch V, et al. Severe chronic insomnia is not associated with higher body mass index. *J Sleep Res* 2015;**24**(5):514–17.
42. Bonnet M, Arand D. Hyperarousal and insomnia: state of the science. *Sleep Med Rev* 2010;**14**(1):9–15.
43. Vgontzas A, Zoumakis M, Papnicolaou D, et al. Chronic insomnia is associated with a shift of interleukin-6 and tumor necrosis factor secretion from nighttime to daytime. *Metabolism* 2002;**51**(7):887–92.
44. Vgontzas A, Liao D, Pejovic S, Calhoun S, et al. Insomnia with objective short sleep duration is associated with type 2 diabetes. *Diabetes Care* 2009;**32**(11):1980–5.
45. Fernandez-Mendoza J, Vgontzas A, Liao D, Shaffer M, et al. Insomnia with objective short sleep duration and incident hypertension. *Hypertension* 2012;**60**(4):929–35.
46. Vgontzas A, Fernandez-Mendoza J, Liao D, Bixler E. Insomnia with objective short sleep duration: the most biologically severe phenotype of the disorder. *Sleep Med Rev* 2013;**17**(4):241–54.
47. Bathgate C, Edinger J, Wyatt J, Krystal A. Objective but not subjective short sleep duration associated with increased risk for hypertension in individuals with insomnia. *Sleep* 2016;**39**(5):1037–45.
48. Seetho I, Wilding J. Sleep-disordered breathing, type 2 diabetes and the metabolic syndrome. *Chron Respir Dis* 2014;**11**(4):257–75.
49. Shrivastava D. Impact of sleep-disordered breathing treatment on upper airway anatomy and physiology. *Sleep Med* 2014;**15**(7):733–41.
50. Vgontzas A, Bixler E, Chrousos G. Sleep apnea is a manifestation of the metabolic syndrome. *Sleep Med Rev* 2005;**9**(3):211–24.
51. Vgontzas A, Tan T, Bixler E, et al. Sleep apnea and sleep disruption in obese patients. *Arch Intern Med* 1994;**154**(15):1705–11.
52. Vgontzas A, Papanicolaou D, Bixler E, Hopper K, et al. Sleep apnea and daytime sleepiness and fatigue: relation to visceral obesity, insulin resistance, and hypercytokinemia. *J Clin Endocrinol Metab* 2000;**85**(3):1151–8.
53. Bixler E, Fernandez-Mendoza J, Liao D, Calhoun S, et al. Natural history of sleep disordered breathing in prepubertal children transitioning to adolescence. *Eur Respir J* 2016;**47**:1402–9.
54. Maas A, Appelman Y. Gender differences in coronary heart disease. *Neth Heart J* 2010;**18**(12):598–602.
55. Gaines J, Vgontzas A, Fernandez-Mendoza J, He F, et al. Increased inflammation from childhood to adolescence predicts sleep apnea in boys: a preliminary study. *Brain Behav Immun* 2017;**64**:259–65.
56. Quan S, Budhiraja R, Clarke D, et al. Impact of treatment with continuous positive airway pressure (CPAP) on weight in obstructive sleep apnea. *J Clin Sleep Med* 2013;**9**(10):989–93.
57. Drager L, Brunoni A, Jenner R, et al. Effects of CPAP on body weight in patients with obstructive sleep apnoea: a meta-analysis of randomised trials. *Thorax* 2015;**70**(3):258–64.
58. Myllyl M, Kurki S, Anttakainen U, et al. High adherence to CPAP treatment does not prevent the continuation of weight gain among severely obese OSAS patients. *J Clin Sleep Med* 2016;**12**(4):519–28.
59. Vgontzas A, Bixler E, Tan T, et al. Obesity without sleep apnea is associated with daytime sleepiness. *Arch Intern Med* 1998;**158**:1333–7.

60. Bixler E, Vgontzas A, Lin H, Calhoun A, et al. Excessive daytime sleepiness in a general population sample: the role of sleep apnea, age, obesity, diabetes, and depression. *J Clin Endocrinol Metab* 2005;**90**(8):4510–15.
61. Fernandez-Mendoza J, Vgontzas A, Kritikou I, et al. Natural history of excessive daytime sleepiness: role of obesity, weight loss, depression, and sleep propensity. *Sleep* 2015;**38**(3):351–60.
62. Yun L, Vgontzas A, Fernandez-Mendoza J, et al. Objective, but not subjective, sleepiness is associated with inflammation in sleep apnea. *Sleep* 2017;**40**(2).
63. Danesh J, Whincup P, Walker M, Lennon L, et al. Low grade inflammation and coronary heart disease: prospective study and updated meta-analyses. *Br Med J* 2000;**321**:199–204.
64. Nishimura S, Manabe I, Nagai R. Adipose tissue inflammation in obesity and metabolic syndrome. *Discov Med* 2009;**8**(41):55–60.
65. Jung U, Choi M. Obesity and its metabolic complications: the role of adipokines and the relationship between obesity, inflammation, insulin resistance, dyslipidemia and nonalcoholic fatty liver disease. *Int J Mol Sci* 2014;**15**(4):6184–223.
66. Ridker P. From C-reactive protein to interleukin-6 to interleukin-1. *Circ Res* 2016;**118**(1):145–56.
67. Gaines J, Vgontzas A, Fernandez-Mendoza J, Calhoun S, et al. Inflammation mediates the association between visceral adiposity and obstructive sleep apnea in adolescents. *Am J Physiol Endocrinol Metab* 2016;**311**(5):851–8.
68. King D, Egan B, Woolson R. Effect of a high-fiber diet vs a fiber-supplemented diet on C-reactive protein level. *Arch Intern Med* 2007;**167**(5):502–6.
69. Nieto-Vazquez I, Fernández-Veledo S, Krämer D, Vila-Bedmar R. Insulin resistance associated to obesity: the link TNF-alpha. *J Metab Dis* 2008;**114**(3):183–94.
70. Fiaschi T, Magherini F, Gamberi T, Modesti P, Modesti A. Adiponectin as a tissue regenerating hormone: more than a metabolic function. *Cell Mol Life Sci* 2014;**71**(10):1917–25.
71. Balsan G, Vieira J, Oliveira A, Portal V. Relationship between adiponectin, obesity and insulin resistance. *Rev Assoc Méd Bras* 2015;**61**(1):72–80.
72. Mutairi S, Mojiminiyi O, Alawi A, et al. Study of leptin and adiponectin as disease markers in subjects with obstructive sleep apnea. *Dis Markers* 2014;706314.
73. Yoshikawa M, Yamauchi M, Fujita Y, et al. The impact of obstructive sleep apnea and nasal CPAP on circulating adiponectin levels. *Lung* 2014;**192**(2):289–95.
74. Vgontzas A, Zoumakis E, Bixler E, et al. Adverse effects of modest sleep restriction on sleepiness, performance, and inflammatory cytokines. *J Clin Endocrinol Metab* 2004;**89**:2119–26.
75. Solomou S, Korbonits M. The role of ghrelin in weight-regulation disorders: implications in clinical practice. *Hormones* 2014;**13**(4):458–75.
76. Wren A, Seal L, Cohen M, Brynes A, et al. Ghrelin enhances appetite and increases food intake in humans. *J Clin Endocrinol Metab* 2001;**86**(12):5992–5.
77. Tamboli R, Breitman I, Marks-Shulman P, Jabbour K, et al. Early weight regain after gastric bypass does not affect insulin sensitivity but is associated with elevated ghrelin. *J Obes (Silver Springs)* 2014;**22**(7):1617–22.
78. Broussard J, Kilkus J, Delebecque F, et al. Elevated ghrelin predicts food intake during experimental sleep restriction. *Obesity (Silver Springs)* 2016;**24**(1):132–8.
79. Paz-Filho G, Mastronardi C, Licinio J. Leptin treatment: facts and expectations. *Metabolism* 2015;**64**(1):146–56.
80. Taheri S, Lin L, Austin D, et al. Short sleep duration is associated with reduced leptin, elevated ghrelin, and increased body mass index. *PLoS Med* 2004;**1**(3):e62.
81. Stern J, Grant A, Thomson C, Tinker L, et al. Short sleep duration is associated with decreased serum leptin, increased energy intake and decreased diet quality in postmenopausal women. *Obesity (Silver Spring)* 2014;**22**(5):E55–61.

82. Pejovic S, Vgontzas A, Basta M, et al. Leptin and hunger levels in young healthy adults after one night of sleep loss. *Sleep Res* 2010;**19**(4):552–8.
83. Bornstein S, Uhlmann K, Haidan A, et al. Evidence for a novel peripheral action of leptin as a metabolic signal to the adrenal gland: leptin inhibits cortisol release directly. *Diabetes* 1997;**46**:1235–8.
84. Chrousos G. Organization and integration of the endocrine system. *Sleep Med Clin* 2007;**2**(2):125–45.
85. Manenschijn L, Schaap L, Schoor N, Pas S, et al. High long-term cortisol levels, measured in scalp hair, are associated with a history of cardiovascular disease. *J Clin Endocrinol Metab* 2013;**98**(5):2078–83.
86. Bitsika V, Sharpley C, Sweeny J, McFarlane J, et al. HPA and SAM axis response as correlates of self- vs parental ratings of anxiety in boys with an autistic disorder. *Physiol Behav* 2014;**127**:1–7.
87. Pejovic S, Basta M, Vgontzas A, Kritikous I, et al. Effects of recovery sleep after one work week of mild sleep restriction on interleukin-6 and cortisol secretion and daytime sleepiness and performance. *Am J Physiol Endocrinol Metab* 2013;**305**(7):E890–6.
88. Fernandez-Mendoza J, He F, LaGrotte C, Vgontzas A, et al. Impact of the metabolic syndrome on mortality is modified by objective short sleep duration. *J Am Heart Assoc* 2017;**6**(5):e005479.
89. Haack M, Sanchez E, Mullington J. Elevated inflammatory markers in response to prolonged sleep restriction are associated with increased pain experience in healthy volunteers. *Sleep* 2007;**30**:1145–52.
90. Irwin M, Wang M, Campomayor C, et al. Sleep deprivation and activation of morning levels of cellular and genomic markers of inflammation. *Arch Intern Med* 2006;**166**:1756–62.
91. Irwin M, Carrillo C, Olmstead R. Sleep loss activates cellular markers of inflammation: sex differences. *Brain Behav Immun* 2010;**24**(1):54–7.
92. Vgontzas A, Pejovic S, Zoumakis E, et al. Daytime napping after a night of sleep loss decreases sleepiness, improves performance, and causes beneficial changes in cortisol and interleukin-6 secretion. *Am J Physiol Endocrinol Metab* 2007;**292**(1):E253–261.

CHAPTER

8

Sleep Disorders and Genes

Eric Murillo-Rodríguez[1,2], Tetsuya Yamamoto[2,3], André Barciela Veras[2,4], Nuno Barbosa Rocha[2,5], Diogo Telles-Correira[2,6], Sérgio Machado[2,7,8,9], Diogo Monteiro[2,10], Henning Budde[2,11,12,13] and Pablo Torterolo[2,14]

[1]Molecular and Integrative Neuroscience Laboratory, School of Medicine, Health Science Division, Universidad Anahuac Mayab, Mérida, Yucatán, México [2]Intercontinental Neuroscience Research Group [3]Graduate School of Technology, Industrial and Social Sciences, Tokushima University, Tokushima, Japan [4]Dom Bosco Catholic University, Campo Grande, Mato Grosso, Brazil [5]Health School, Polytechnic Institute of Porto, Porto, Portugal [6]Faculty of Medicine, University of Lisbon, Lisbon, Portugal [7]Laboratory of Panic and Respiration, Institute of Psychiatry, Federal University of Rio de Janeiro, Rio de Janeiro, Brazil [8]Salgado de Oliveira University, Rio de Janeiro, Brazil [9]Physical Activity Neuroscience Laboratory, Physical Activity Sciences Postgraduate Program, Salgado de Oliveira University (UNIVERSO), Rio de Janeiro, Brazil [10]Sport Science School, Polytechnic Institute of Santarém, Santarém, Portugal [11]Faculty of Human Sciences, Medical School Hamburg, Hamburg, Germany [12]Physical Activity, Physical Education, Health and Sport Research Centre (PAPESH), Sports Science Department, School of Science and Engineering, Reykjavik University, Reykjavik, Iceland [13]Lithuanian Sports University, Kaunas, Lithuania [14]Laboratory of Sleep Neurobiology, Department of Physiology, School of Medicine, Universidad de la República, Montevideo, Uruguay

Molecular Sleep—Wake Cycle
DOI: https://doi.org/10.1016/B978-0-12-816430-3.00008-7

OUTLINE

8.1 INTRODUCTION

Diverse neurobiological elements such as neurochemical, neuroanatomical, and genetic modulate the sleep–wake cycle. In addition, it has been described as the link between gene expression and sleep disturbances. Among the most frequent sleep alterations are insomnia and excessive daytime sleepiness. However, there are additional kinds of sleep disturbances such as obstructive sleep apnea, narcolepsy, advanced sleep phase syndrome, restless legs syndrome, among others. Importantly, current evidence has demonstrated the genetic basis of multiple sleep disorders. Thus, it is likely that genetic phenomena contribute to the control of the sleep–wake cycle. Moreover, new therapeutical approaches aimed at managing sleep disorders should consider the genetic influence in the onset of pathologies related to sleep.

8.2 THE SLEEP–WAKE CYCLE: AN OVERVIEW

In humans, mammals, and birds, three behavioral states can be readily distinguished: wakefulness (W), slow wave sleep (SWS), and rapid eye movement sleep (REMS). Polysomnography is the basic tool used to differentiate these states (Fig. 8.1). It consists of the simultaneous recording of various physiological parameters such as the electroencephalogram (EEG), the electromyogram (EMG), and eye movements.

8.2.1 Wakefulness

During W, there is an optimal interaction with the environment that allows the development of various behaviors necessary for survival. In humans, W is accompanied by awareness (consciousness) of the environment and internally generated stimuli such as hunger and thirst. The EEG recording of W is defined by the presence of high frequency (above 15 Hz) and low amplitude oscillations determined by the activity of thalamic and cortical neurons. In spite of this, during relaxed W with the eyes closed, regular alpha (8–12 Hz) activity predominates in the occipital region.

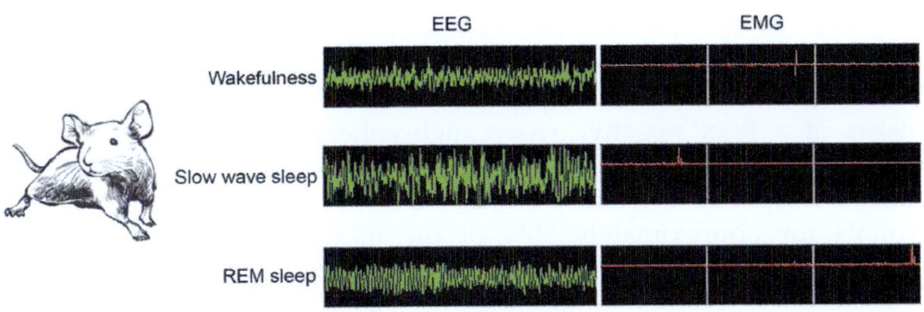

FIGURE 8.1 Polysomnographic features of the sleep–wake cycle. The states of vigilance comprise wakefulness, slow wave sleep, and rapid eye movement (REM) sleep by electroencephalogram (EEG) and electromyogram (EMG) traces. Wakefulness shows EEG patterns of rapid (alpha) activity (8–13 Hz) and higher activity is recorded in the EMG. The slow wave sleep state displays EEG spindle activity as well as a low voltage, including K-complexes and waves over 100 μV of amplitude and less than 2 Hz frequency accompanied with EMG reduced activity. REM sleep displays in the EEG low voltage mixed frequencies (2–7 Hz) as well as muscle twitches recorded in the EMG.

W-promoting neuronal networks, known as activating systems, are located in the reticular formation of the upper brainstem, posterolateral hypothalamus, and basal forebrain.[1] These neurons are critical for the generation and maintenance of W; lesions on these areas generate a deficit in the maintenance of W, including coma.[2] It is also important to note that narcolepsy, a paradigmatic hypersomnia, occurs because of a degeneration of neurons of the posterolateral region of the hypothalamus that synthesize and utilize the neuropeptides hypocretin 1 and 2 (also called orexin A and B) as neuromodulators. The hypocretinergic neurons promote W, and are part of the activating system.[3,4] Neurons that constitute the activating system inhibit SWS promoting neuronal networks. In fact, waking and SWS promoting neuronal networks have reciprocal inhibition between them.[1]

8.2.2 Slow Wave Sleep

During sleep, there is a marked decrease in the interaction with the environment, an increase of the threshold for the reaction to external stimuli, and a decrease in somatomotor activity. Furthermore, animals adopt a distinctive position to conserve heat.

In healthy adults, the transition from W to sleep is into SWS sleep. In humans, three SWS phases are distinguished according to the depth of the state: S1, S2, and S3. From W, normal adults enter in S1 or light SWS. This transitional state present low-voltage, mixed frequency waves (2–7 Hz range) in the EEG. The S2 is characterized by the presence of sleep spindles and K-complexes in the EEG. Sleep spindles are "spindle-like" bursts of oscillatory activity of 11–15 Hz with a duration of 0.5–2 seconds; the intervals between sleep spindles in human subjects are 3–5 seconds.[5] These events are generated by the interplay of the reticular thalamic nucleus with other thalamic nuclei. The K-complex consists of a brief negative sharp high-voltage peak (usually greater than 100 μV), followed by a slower positive complex and final negative peak. K-complexes are often associated with sleep spindles. The S2 accounts for approximately 50% of the total sleep time during the night.[6] The S3 is characterized by low frequency (0.5–4 Hz, delta band of frequencies) of high amplitude ($>75\,\mu$V) waves in the EEG. S3 accounts for 12.5%–20% of the total sleep time.[6]

Tonic parasympathetic activity increases during SWS determining characteristic changes in visceral activity, such as a pronounced decrease in heart rate.[7] In the deeper stages of SWS, cognitive activity (i.e., dreaming) is minimal.[8–10] In the preoptic area of the hypothalamus are the main SWS promoting neuronal networks. Most of these neurons are GABAergic and inhibit the activating systems to generate SWS.[11] In fact, pioneer studies by Von Economo showed that lesions of this area generate severe insomnia.[12]

8.2.3 Rapid Eye Movement Sleep

REMS occurs periodically, and in normal adults is always preceded by SWS. REMS is a deep sleep stage although the EEG is similar to that of W (high frequency and low amplitude oscillations); hence, it is also called "paradoxical" sleep. REMS is also distinguished by fast, saccadic rapid eye movements (REMs) and muscle atonia as evidenced in the EMG channel. Muscle atonia is produced mainly by postsynaptic hyperpolarization of the spinal and brainstem motoneurons.[13] This REMS atonia is absent or reduced in REMS behavioral disorder, a parasomnia characterized by patients acting out their dreams.[14] Phasic changes in autonomic activity, which generates irregular visceral activity (such as increases and decreases in blood pressure), are also associated with REMS.[15] This state is also characterized by a lack of homeostatic control of body temperature.[16]

Cognitive activities do not just occur during W. Dreams, which occur more prominently during REMS,[8–10] are considered a special kind of cognitive activity or protoconsciousness.[17] REM sleep dreams are characterized by their vividness, single-mindedness, bizarreness, and loss of voluntary control over the plot. Attention is unstable and rigidly focused, facts and reality are not checked, violation of physical laws and bizarreness are passively accepted, contextual congruence is distorted, and time is altered.[17–19]

Dreams and psychosis share important characteristics such as internal perceptions independent of external stimulation with a general lack of criticism. In addition, REMS shares neurophysiological, and neurochemical characteristics with psychosis; because of this fact, it is considered a natural model of this condition.[20–22]

REMS occupies 20%–25% of total sleep in human adults, and appears about ~90 minutes after sleep onset (REMS latency).[6] A shortened REMS latency (the interval between the sleep onset and the appearance of the first REMS episode) is a biological marker of primary depression.[23] It is also considered to be a clinically significant pathological feature in narcolepsy.[24]

Sleep during the night shows four to five SWS-REMS cycles (the period from sleep onset to the end of the first REMS episode or the period from the end of a REMS episode to the subsequent REMS episode). The average length of human sleep cycles is about 90–110 minutes.[25]

Neuronal networks of the pontomesencephalic reticular formation are necessary and sufficient for REMS generation.[26] These neuronal networks are also involved in the control of W (constituting the ascending reticular activating system). However, the activity of these neurons differs between W and REMS. More recently, neurons of the lateral hypothalamus and incertohypothalamic area that synthesize and employ the neuropeptide melanin-concentrating hormone (MCH) as a neuromodulator, have been involved both in the generation of SWS and REMS.[27,28] In addition, malfunctioning of the MCHergic system may be involved in the pathophysiology of depression.[29] MCH antagonists are considered promising antidepressive and anxiolytic drugs.[30]

8.2.4 Circadian and Homeostatic Regulation of Sleep

The physiological transition between W and sleep is regulated by a circadian and a homeostatic component.[31] Like all circadian rhythms, sleep and W are regulated by commands from the suprachiasmatic nucleus (SCN) of the hypothalamus. The SCN receives photic information directly from the retina, and regulates the activity of both the hypnogenic and activating systems.[32] Furthermore, through indirect modulation of the sympathetic system, the SCN regulates the release of melatonin from the pineal gland during the night.[33] Melatonin has a weak sleep-promoting effect.

The homeostatic component also regulates the sleep–wake cycle; that is, prolonged W facilitates the generation of sleep. Different lines of research have shown that sleep-promoting substances, including adenosine, are released and accumulated during W.[34,35] Adenosine promotes sleep by inhibiting the activating systems and stimulating SWS promoting systems. Of interest, caffeine promotes W by blocking the receptors for adenosine.[36]

Sleep is also deeply modified with age. In the newborn, transition from W to sleep is often accomplishes through REMS (called active sleep in newborns). The sleep cycle (alternation of SWS-REMS) period in newborn is of 50–60 minutes and the circadian organization is not present.[37,38] In addition, a shortening in sleep duration, mainly related to the reduction in S3 and REMS, is associated with aging.[25]

8.3 BASIS OF GENE EXPRESSION

It is widely accepted that genes are the database contained inside the nucleus of all cells, and are responsible for physical and functional characteristics of subjects. Since every gene contains a particular set of molecular instructions for codifying a specific protein, this process is known as gene expression.[39,40] Remarkably, there is a natural variation in human gene expression. Every single person possesses two copies of each gene (one comes from each parent) and in turn, each human being has their own genetic information, which was inherited from their progenitors. Thus, the physical characteristics, physiology, and even predisposition to certain pathologic conditions are also coded in genes. However, it is worthy to mention that not all genes are active at all times. It seems that some genes respond only when they interact with the environment, remaining active only for a limited time window.[41] Since some genes have been linked with sleep disorders, this review provides a broad understanding of the present knowledge regarding the sleep disorders and gene expression.

8.4 GENE EXPRESSION IN PATHOLOGICAL CONDITIONS

As mentioned previously, genes contain molecular instructions for codifying a specific protein, including some aberrant features. Clinically, we can interpret this molecular event as pathology. Most known cases of diseases linked to gene expression are cancer, diabetes, neurodegenerative disorders, obesity, and many others.[42] For example, it has been reported that the protein Semaphorin-3E (Sema3E) shows an overexpression in human pancreatic cancer. In addition, Sema3E has been suggested as a contributor to tumor progression and metastasis, since it has been detected in pancreatic cancer cells inducing cellular proliferation. In opposition, experimental Sema3E knockout animals display a suppression of cancer cellular proliferation and a reduction in tumor size.[43] Another example of the role of nuclear elements related to diseases is mRNA. In this regard, osteoporosis is a common disease in the aging population, and it has been suggested that this disturbance is related to the normal decalcification rate. The origin of this health problem has been associated to causes such as deficiency in vitamin D, calcium, addiction to alcohol, as well as gene expression. Recent evidence has demonstrated that circulating mRNA levels in postmenopausal subjects could be used as potential biomarkers for osteoporosis.[44] Given the complex relationship between health disturbances and gene expression, the extensiveness of this field is not a surprise.

8.5 SLEEP DISORDERS LINKED TO GENE EXPRESSION

8.5.1 Insomnia

In previous chapters of the current book, several sleep disturbances have been described. Thus, we will avoid redundancy in information.

In the case of insomnia, this sleep disturbance is characterized by the difficulty to initiate or maintain sleep. Regarding the genetic basis of insomnia, several pieces of evidence have reported the critical role of multiple circadian genes, such as *CLOCK*, *PER*, *Timeless*, etc., in subjects who present insomnia as well as mood disorders.[45–51] Additional examples of the relationship between specific gene expression and insomnia include the data reported about genes that are related to the duration of a precise state of vigilance. In this case, *Per2* and *Per3* activity have shown higher patterns of expression and insomnia.[49,50]

Additional genes have been added to the list of genes related to insomnia. For example, adenosine, GABA, serotonin transporter polymorphic region (*5-HTTLPR*), and orexin/hypocretin show higher pattern of expression in insomniac conditions.[45] To further complicate the scenario, several reports have suggested the involvement of additional wake-related neurotransmitter systems, including dopaminergic system genes [i.e., dopamine transporter 1 (*DAT1*), catecholamine-*O*-methyltransferase (*COMT*), dopamine receptor D4 (*DRD4*)[45,52]]. Current genome-wide association studies (GWAS) have reported novel genes that seems to be associated to insomnia, such as the *ABCC9* gene (rs11046209).[52–55]

8.5.2 Obstructive Sleep Apnea

The clinical feature of repeated cessation or attenuation of breathing (called apneas and hypopneas, respectively) during sleep is known as obstructive sleep apnea (OSA). This sleep disorder is also linked with several health dysfunctions such as hypertension, obesity, heart failure, among others.[56] The first-line treatment given to subjects who present with OSA comprises the use of a piece of equipment that provides air pressure, known as continuous positive airway pressure or CPAP.[56,57]

The molecular mechanisms underlying OSA remain unclear. However, there is evidence that suggests the involvement of genes. In this regard, GWAS in 12,558 Hispanic subjects found two novel loci at genome level with significance for apnea-hypopnea index.[58] In addition, it has been proposed that genes related with inflammation might be linked to OSA. Thus, inflammatory factors, including IL-6, IL-8, and TNF-α, have been found highly expressed in patients with OSA.[59,60] Additional gene candidates for OSA promotion are the 5-hydroxytryptamine receptor 2A (*5-HTR2A*) and interleukin-6 (*IL-6*).[61,62]

8.5.3 Narcolepsy

The major symptoms of narcolepsy comprise excessive daytime sleepiness and cataplexy. Regarding the genetic basis of this sleep disorder, it has been suggested that the human leukocyte antigen (HLA)-DRB1 × 15:01-DQB1 × 06:02 haplotype seems to be linked with narcolepsy.[63–65] In addition, GWAS in 525,196 single nucleotide polymorphisms (SNPs) have been identified outside of the HLA region and associated to narcolepsy.[66] Recent studies have shown that SNP in the promoter region of chemokine (C–C motif) receptor 1 (CCR1) is also linked to narcolepsy.

8.5.4 Advanced Sleep Phase Syndrome

The Earth rotates once in about 24 hours with respect to the Sun, but once every 23 hours, 56 minutes, and 4 seconds approximately with respect to the stars. Thus, this cycle, called the *circadian rhythm*, has established a remarkable influence in the neurobiological activity of several species, including humans. In this regard, the circadian rhythm has also evidenced the existence of pathological circadian-related issues, such as sleep disturbances.[67–71] For example, the advanced sleep phase syndrome is characterized by subjects' complaint of early evening bedtimes or early morning awakenings. In this case, this abnormal feature of sleep onset is called advanced sleep phase syndrome. This sleep disturbance is evident when social constraints force the patients to late bedtimes.[72,73]

Sleep control under the influence of the circadian rhythm responds to specific genes. In this regard, approximately 10% of the population are homozygous for the 5-repeat allele (*PER3*(5/5)) of a variable number tandem repeated polymorphism in the clock gene *Per3*. Importantly, *PER3*(5/5) has been linked to morning preference, while homozygosity for the 4-repeat allele [*PER3*(4/4)] seems to be associated with evening preference.[74]

Genetic studies have contributed to the understanding of the molecular basis of circadian rhythms and associated sleep disorders. In this regard, more than 20 genes have been characterized, including *PER 1, 2,* and *3*; the cryptochrome genes 1 and 2; and brain and muscle arnt-like protein-1 [Bmal1/Aryl hydrocarbon receptor nuclear translocator-like (ARNTL1),[75,76]].

It is worthy to mention that diurnal preference seems also to be the result of activation of specific genes. For example, diurnal preference and a polymorphism in *Per3* or polymorphism in aryl hydrocarbon receptor nuclear translocator-like 2 (*ARNTL2*) have been described.[77] In addition, significant gene-associated loci with morningness has been suggested.[78] Taken together, we can conclude that genes with circadian expression play a key role in modulation of sleep onset as well as in circadian-related sleep disturbances.[79–83]

8.5.5 Restless Legs Syndrome

Restless legs syndrome is a pathological condition characterized by the urge to move the legs during W, and is associated with periodic leg movements during sleep.

The substantia nigra, a critical brain area for modulating movement, is composed of dopaminergic neurons that project to the striatum; low iron concentration within this area seems to be involved in the genesis of this sleep disorder.[84,85] Although significant advances have been achieved in the understanding of restless legs syndrome, the genetics of this sleep disorder are still unknown. Current evidence suggests that genes such as IL-β or BTBD9 participate in the onset of the motor disorder.[86,87] In addition, since this sleep disturbance has familial aggregation, GWAS have identified SNPs linked to the restless legs syndrome, including a SNP at loci MEIS1.[88–90] Indeed, more research is needed to clarify the role of genes that are participating in the onset of restless legs syndrome.[84–91]

8.6 COULD WE EDIT GENES RELATED TO SLEEP DISORDERS? THE CASE OF CRISPR-CAS9

Recently, the gene editing technique known as Clustered Regularly Interspaced Short Palindromic Repeats (CRISPR) has been reported. Briefly, CRISPR uses a guide RNA to send biological "scissors"—usually the CRISPR-associated protein,

Cas9 is used at a precise spot to cut and make a substitution in the genome. As one can imagine, CRISPR-Cas9 provides the technology to manipulate genomic sequences, focusing on human diseases, such as HIV and cancer.[92–96] Moreover, CRISPR-Cas9 might be also considered for editing genes in sleep disorders, such as insomnia, OSA, narcolepsy, advanced sleep phase syndrome, or restless legs syndrome[97,98] (Fig. 8.2). Indeed, it is needed to explore of the use of CRISPR-Cas9 in animal models of sleep disorders before considering it a "bench-to-bedside" approach.[99,100]

8.7 CONCLUSIONS

The sleep–wake cycle is a physiological phenomenon that displays, as many other functions, aberrant features that have been characterized as sleep disturbances. Within the family of sleep disorders, insomnia, OSA, narcolepsy, advanced sleep phase syndrome, and

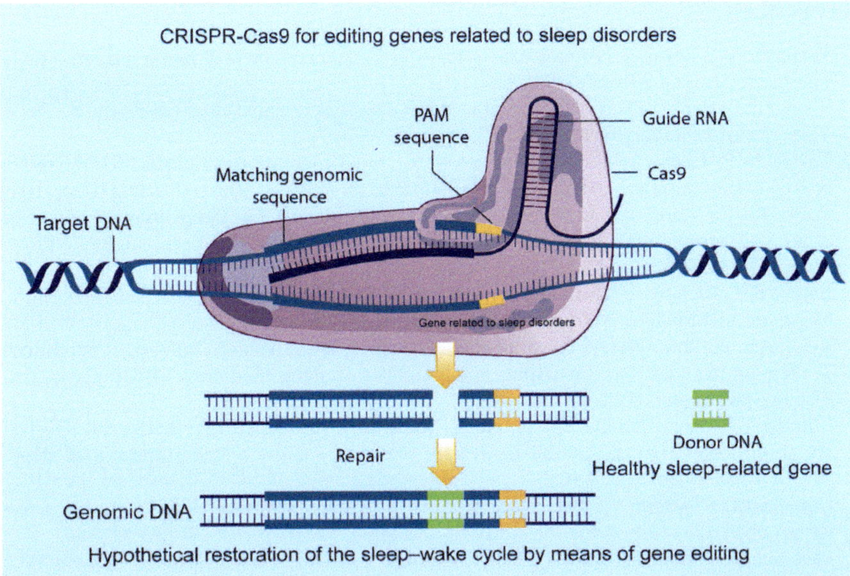

FIGURE 8.2 Putative model of hypothetical use of CRISPR/Cas9 for editing genes related to sleep disorders. By molecular manipulation of genes linked to sleep disturbances, a novel therapeutical approach might represent an efficient treatment for insomnia, OSA, narcolepsy, advanced sleep phase syndrome, restless legs syndrome, or other sleep disturbances.

restless legs syndrome are some of the unhealthy features of sleep. Research aimed to understand the origin of sleep disturbances has included the studies of gene expression linked to those sleep pathologies. In this regard, identifying the genes related to sleep disturbances has led to significant achievements in the sleep medicine area. Currently, the modern molecular technique CRISPR-Cas9 might be considered to editing genes and its putative use is highly tempting to regulate or eliminate the sleep disturbances. Indeed, further studies are required to describe the potential use of CRISPR-Cas9 in the editing of genes related to sleep disorders.

Acknowledgments

This work was supported by The University of California Institute for Mexico and the United States (UC MEXUS) and Consejo Nacional de Ciencia y Tecnología (CONACyT; Grant# CN-17–19) and Escuela de Medicina, Universidad Anáhuac Mayab (Grant: PresInvEMR2017) given to E. M-R.

References

1. Torterolo P, Vanini G. Nuevos conceptos sobre la generación y el mantenimiento de la vigilia. *Rev Neurol* 2010;**50**:747–58.
2. Posner J, Saper CB, Schiff ND, Plum F. *The diagnosis of stupor and coma*. 4th ed. New York: Oxford University Press; 2007.
3. Torterolo P, Chase MH. The hypocretins (orexins) mediate the "phasic" components of REM sleep: a new hypothesis. *Sleep Sci* 2014;**7**:19–29.
4. Torterolo P, Monti J, Pandi-Perumal SR. Role of orexin on sleep: interactions with other neurotransmitter systems. In: Sakurai T, Monti J, Pandi-Perumal SR, editors. *Orexin and sleep*. Springer; 2015. p. 181–202.
5. Evans BM, Richardson NE. Demonstration of a 3-5s periodicity between the spindle bursts in NREM sleep in man. *J Sleep Res* 1995;**4**:196–7.
6. Keenan S, Hirshkowitz M. Monitoring and staging human sleep. In: Kryger MH, Roth T, Dement WC, editors. Principles and practices of sleep medicine. Philadelphia, PA: Elsevier-Saunders; 2011. p. 1602–9.
7. Brando V, Castro-Zaballa S, Falconi A, Torterolo P, Migliaro ER. Statistical, spectral and non-linear analysis of the heart rate variability during wakefulness and sleep. *Arch Ital Biol* 2014;**152**:32–46.
8. Dement W, Kleitman N. The relation of eye movements during sleep to dream activity: an objective method for the study of dreaming. *J Exp Psychol* 1957;**53**:339–46.
9. Pace-Schott E. The neurobiology of dreaming. In: Kryger MH, Roth T, Dement WC, editors. *Principles and practices of sleep medicine*. Philadelphia, PA: Elsevier-Saunders; 2011. p. 563–75.
10. Siclari F, Baird B, Perogamvros L, Bernardi G, LaRocque JJ, Riedner B, et al. The neural correlates of dreaming. *Nat Neurosci* 2017;**20**:872–8.
11. Scammell TE, Arrigoni E, Lipton JO. Neural circuitry of wakefulness and sleep. *Neuron* 2017;**93**:747–65.
12. Von Economo C. Sleep as a problem of localization. *J Nerv Ment Dis* 1930;**71**:249–59.
13. Chase MH. Motor control during sleep and wakefulness: clarifying controversies and resolving paradoxes. *Sleep Med Rev* 2013;**17**:299–312.
14. Mahowald MW, Schneck C. REM sleep parasomnias. In: Kryger MH, Roth T, Dement WC, editors. *Principles and practices of sleep medicine*. Philadelphia, PA: Elsevier-Saunders; 2011. p. 1083–97.
15. Amici R, Cerri M, Parmeggiani PL. *Overview of physiological processes during sleep. Encyclopedia of sleep*. Elsevier; 2013.
16. Parmeggiani PL. Thermoregulation and sleep. *Front Biosci* 2003;**8**:s557–67.
17. Hobson JA. REM sleep and dreaming: towards a theory of protoconsciousness. *Nat Rev Neurosci* 2009;**10**:803–13.
18. Rechtschaffen A. The single-mindedness and isolation of dreams. *Sleep* 1978;**1**:97–109.
19. Nir Y, Tononi G. Dreaming and the brain: from phenomenology to neurophysiology. *Trends Cogn Sci* 2010;**14**:88–100.
20. Naiman R. Dreamless: the silent epidemic of REM sleep loss. *Ann NY Acad Sci* 2017;**1406**(1):77–85.
21. Gottesmann C. The dreaming sleep stage: a new neurobiological model of schizophrenia? *Neuroscience* 2006;**140**:1105–15.
22. Gottesmann C, Gottesman I. The neurobiological characteristics of rapid eye movement (REM) sleep are candidate endophenotypes of depression, schizophrenia, mental retardation and dementia. *Prog Neurobiol* 2007;**81**:237–50.
23. Palagini L, Baglioni C, Ciapparelli A, Gemignani A, Riemann D. REM sleep dysregulation in depression: state of the art. *Sleep Med Rev* 2013;**17**(5):377–90.

24. Guilleminault C, Fromherz S. Narcolepsy: diagnosis and management. In: Kryger MH, Roth T, Dement WC, editors. *Principles and practices of sleep medicine*. Philadelphia, PA: Saunders; 2011. p. 957–68.
25. Carskadon MA, Dement W. Normal human sleep: an overview. In: Kryger MH, Roth T, Dement W, editors. *Principles and practices of sleep medicine*. Philadelphia, PA: Elsevier-Saunders; 2011. p. 16–26.
26. Siegel JM. REM sleep. In: Kryger MH, Roth T, Dement WC, editors. *Principles and practices of sleep medicine*. Philadelphia, PA: Elsevier-Saunders; 2011. p. 92–111.
27. Torterolo P, Lagos P, Monti JM. Melanin-concentrating hormone (MCH): a new sleep factor? *Front Neurol* 2011;**2**:1–12.
28. Monti JM, Torterolo P, Lagos P. Melanin-concentrating hormone control of sleep-wake behavior. *Sleep Med Rev* 2013;**17**:293–8.
29. Torterolo P, Scorza C, Lagos P, Urbanavicius J, Benedetto L, Pascovich C, et al. Melanin-concentrating hormone (MCH): role in REM sleep and depression. *Front Neurosci* 2015;**9**:475.
30. Chaki S. MCH receptor 1 antagonists: antidepressant/anxiolytic potential in animal models. In: Pandi Perumal SR, Torterolo P, Monti J, editors. *Melanin-concentrating hormone and sleep*. Switzerland: Springer; 2018. p. 207–25.
31. Borbely AA. A two process model of sleep regulation. *Hum Neurobiol* 1982;**1**:195–204.
32. Mistlberger RE. Circadian regulation of sleep in mammals: role of the suprachiasmatic nucleus. *Brain Res Brain Res Rev* 2005;**49**:429–54.
33. Pandi-Perumal SR, Trakht I, Srinivasan V, Spence DW, Maestroni GJ, Zisapel N, et al. Physiological effects of melatonin: role of melatonin receptors and signal transduction pathways. *Prog Neurobiol* 2008;**85**:335–53.
34. Basheer R, Strecker RE, Thakkar MM, McCarley RW. Adenosine and sleep-wake regulation. *Prog Neurobiol* 2004;**73**:379–96.
35. Huang ZL, Urade Y, Hayaishi O. The role of adenosine in the regulation of sleep. *Curr Top Med Chem* 2011;**11**:1047–57.
36. Nishino S, Mignot E. Wake-promoting medications: basic mechanisms and pharmacology. In: Kryger MH, Roth T, Dement WC, editors. *Principles and practices of sleep medicine*. Philadelphia, PA: Elsevier-Saunders; 2005. p. 468–83.
37. Roffwarg HP, Muzio JN, Dement WC. Ontogenetic development of the human sleep-dream cycle. *Science* 1966;**152**:604–19.
38. Peirano PD, Algarin CR. Sleep in brain development. *Biol Res* 2007;**40**:471–8.
39. Chatterjee S, Ahituv N. Gene regulatory elements, major drivers of human disease. *Annu Rev Genomics Hum Genet* 2017;**18**:45–63.
40. Petit F, Sears KE, Ahituv N. Limb development: a paradigm of gene regulation. *Nat Rev Genet* 2017;**18**:245–58.
41. Lusk CP, King MC. The nucleus: keeping it together by keeping it apart. *Curr Opin Cell Biol* 2017;**44**:44–50.
42. Barbagallo I, Li Volti G, Galvano F, Tettamanti G, Pluchinotta FR, Bergante S, et al. Diabetic human adipose tissue derived mesenchymal stem cells fail to differentiate in functional adipocytes. *Exp Biol Med* 2016;**242**:1079–85.
43. Yong LK, Lai S, Liang Z, Poteet E, Chen F, van Buren G, et al. Overexpression of Semaphorin-3E enhances pancreatic cancer cell growth and associates with poor patient survival. *Oncotarget* 2016;**7**:87431–48.
44. Chen J, Li K, Pang Q, Yang C, Zhang H, Wu F, et al. Identification of suitable reference gene and biomarkers of serum miRNAs for osteoporosis. *Sci Rep* 2016;**6**:36347.
45. Lind MJ, Gehrman PR. Genetic pathways to insomnia. *Brain Sci* 2010;**6**:E64.
46. Serretti A, Benedetti F, Mandelli L, Lorenzi C, Pirovano A, Colombo C, et al. Genetic dissection of psychopathological symptoms: insomnia in mood disorders and CLOCK gene polymorphism. *Am J Med Gen* 2010;**121B**:35–8.

47. Serretti A, Gaspar-Barba E, Calati R, Cruz-Fuentes CS, Gomez-Sanchez A, Perez-Molina A, et al. 3111T/C clock gene polymorphism is not associated with sleep disturbances in untreated depressed patients. *Chronobiol Int* 2003;**27**:265–77.
48. Utge SJ, Soronen P, Loukola A, Kronholm E, Ollila HM, Pirkola S, et al. Systematic analysis of circadian genes in a population-based sample reveals association of TIMELESS with depression and sleep disturbance. *PLoS One* 2010;**5**:e9259.
49. Brower KJ, Wojnar M, Sliwerska E, Armitage R, Burmeister M. PER3 polymorphism and insomnia severity in alcohol dependence. *Sleep* 2012;**35**:571–7.
50. Li J, Huang C, Lan Y, Wang Y. A cross-sectional study on the relationships among the polymorphism of period2 gene, work stress, and insomnia. *Sleep Breath* 2015;**19**:1399–406.
51. Mendlewicz J. Disruption of the circadian timing systems: molecular mechanisms in mood disorders. *CNS Drugs* 2009;**23**:15–26.
52. Ban HJ, Kim SC, Seo J, Kang HB, Choi JK. Genetic and metabolic characterization of insomnia. *PLoS One* 2011;**6**:e18455.
53. Byrne EM, Gehrman PR, Medland SE, Nyholt DR, Heath AC, Madden PA, et al. Chronogen Consortium. A genome-wide association study of sleep habits and insomnia. *Am J Med Gen* 2013;**162**:439–51.
54. Ollila HM, Kettunen J, Pietiläinen O, Aho V, Silander K, Kronholm E, et al. Genome-wide association study of sleep duration in the Finnish population. *J Sleep Res* 2016;**23**:609–18.
55. Spada J, Scholz M, Kirsten H, Hensch T, Horn K, Jawinski P, et al. Genome-wide association analysis of actigraphic sleep phenotypes in the LIFE Adult Study. *J Sleep Res* 2016;**25**:690–701.
56. Subramani Y, Singh M, Wong J, Kushida CA, Malhotra A, Chung F. Understanding phenotypes of obstructive sleep apnea: applications in anesthesia, surgery, and perioperative medicine. *Anesth Analg* 2017;**124**:179–91.
57. Patel N, Donahue C, Shenoy A, Patel A, El-Sherif N. Obstructive sleep apnea and arrhythmia: a systemic review. *Int J Cardiol* 2017;**228**:967–70.
58. Cade BE, Chen H, Stilp AM, Gleason KJ, Sofer T, Ancoli-Israel S, et al. Genetic associations with obstructive sleep apnea traits in Hispanic/Latino Americans. *Am J Respir Crit Care Med* 2016;**194**:886–97.
59. Huang YS, Guilleminault C, Hwang M, Cheng C, Lin CH, Li HY, et al. Inflammatory cytokines in pediatric obstructive sleep apnea. *Med* 2016;**95**:e4944.
60. Hirotsu C, Albuquerque RG, Nogueira H, Hachul H, Bittencourt L, Tufik S, et al. The relationship between sleep apnea, metabolic dysfunction and inflammation: the gender influence. *Brain Behav Immun* 2017;**59**:211–18.
61. Wu W, Li Z, Tang T, Wu J, Liu F, Gu L. *5-HTR2A* and *IL-6* polymorphisms and obstructive sleep apnea-hypopnea syndrome. *Biomed Rep* 2016;**4**:203–8.
62. de Lima FF, Mazzotti DR, Tufik S, Bittencourt L. The role inflammatory response genes in obstructive sleep apnea syndrome: a review. *Sleep Breath* 2016;**20**:331–8.
63. Dong XS, Ma SF, Cao CW, Li J, An P, Zhao L, et al. Hypocretin (orexin) neuropeptide precursor gene, HCRT, polymorphisms in early-onset narcolepsy with cataplexy. *Sleep Med* 2013;**14**:482–7.
64. Han F, Lin L, Schormair B, Pizza F, Plazzi G, Ollila HM, et al. HLA DQB1*06:02 negative narcolepsy with hypocretin/orexin deficiency. *Sleep* 2014;**37**:1601–8.
65. Miyagawa T, Toyoda H, Hirataka A, Kanbayashi T, Imanishi A, Sagawa Y, et al. New susceptibility variants to narcolepsy identified in HLA class II region. *Human Mol Gen* 2015;**24**:891–8.
66. Toyoda H, Miyagawa T, Koike A, Kanbayashi T, Imanishi A, Sagawa Y, et al. A polymorphism in CCR1/CCR3 is associated with narcolepsy. *Brain Behav Immun* 2015;**49**:148–55.

67. Johansson AS, Owe-Larsson B, Hetta J, Lundkvist GB. Altered circadian clock gene expression in patients with schizophrenia. *Schizophrenia Res* 2016;**174**:17–23.
68. Luo W, Ma S, Yang Y, Zhang D, Zhang Q, Liu Y, et al. TFEB regulates PER3 expression via glucose-dependent effects on CLOCK/BMAL1. *Int J Biochem Cell Biol* 2016;**78**:31–42.
69. Panda S. Circadian physiology of metabolism. *Science* 2016;**354**:1008–15.
70. Riddle M, Mezias E, Foley D, LeSauter J, Silver R. Differential localization of PER1 and PER2 in the brain master circadian clock. *Eur J Neurosci* 2016;**45**:1357–67.
71. Videnovic A, Willis GL. Circadian system-a novel diagnostic and therapeutic target in Parkinson's disease? *Mov Dis* 2016;**31**:260–9.
72. Potter GD, Skene DJ, Arendt J, Cade JE, Grant PJ, Hardie LJ. Circadian rhythm and sleep disruption: causes, metabolic consequences, and countermeasures. *Endocrin Rev* 2016;**37**:584–608.
73. van Maanen A, Meijer AM, van der Heijden KB, Oort FJ. The effects of light therapy on sleep problems: a systematic review and meta-analysis. *Sleep Med Rev* 2016;**29**:52–62.
74. Dijk DJ, Archer SN. PERIOD3, circadian phenotypes, and sleep homeostasis. *Sleep Med Rev* 2010;**14**:151–60.
75. Nordgren A. Genes, body clocks and prevention of sleep problems. *Med Health Care Philos* 2016;**19**:569–79.
76. Turek FW. Circadian clocks: not your grandfather's clock. *Science* 2016;**354**:992–3.
77. Parsons MJ, Lester KJ, Barclay NL, Archer SN, Nolan PM, Eley TC, et al. Polymorphisms in the circadian expressed genes PER3 and ARNTL2 are associated with diurnal preference and GNβ3 with sleep measures. *Sleep Res* 2014;**23**:595–604.
78. Hu Y, Shmygelska A, Tran D, Eriksson N, Tung JY, Hinds DA. GWAS of 89,283 individuals identifies genetic variants associated with self-reporting of being a morning person. *Nat Commun* 2016;**7**:10448.
79. Amin N, Allebrandt KV, van der Spek A, Müller-Myhsok B, Hek K, Teder-Laving M, et al. Genetic variants in RBFOX3 are associated with sleep latency. *Eur J Human Genet* 2016;**24**:1488–95.
80. Cade BE, Gottlieb DJ, Lauderdale DS, Bennett DA, Buchman AS, Buxbaum SG, et al. Common variants in DRD2 are associated with sleep duration: the CARe consortium. *Human Mol Genet* 2016;**25**:167–79.
81. Chang AM, Bjonnes AC, Aeschbach D. Circadian gene variants influence sleep and the sleep electroencephalogram in humans. *Chronobiol Int* 2016;**33**:561–73.
82. Dall'Ara I, Ghirotto S, Ingusci S, Bagarolo G, Bertolucci C, Barbujani G. Demographic history and adaptation account for clock gene diversity in humans. *Heredity* 2016;**117**:165–72.
83. Truong KK, Lam MT, Grandner MA, Sassoon CS, Malhotra A. Timing matters: circadian rhythm in sepsis, obstructive lung disease, obstructive sleep apnea, and cancer. *Ann Am Thorac Soc* 2016;**13**:1144–54.
84. Rizzo G, Li X, Galantucci S, Filippi M, Cho YW. Brain imaging and networks in restless legs syndrome. *Sleep Med* 2016;**31**:39–48.
85. Scherer JS, Combs SA, Brennan F. Sleep disorders, restless legs syndrome, and uremic pruritus: diagnosis and treatment of common symptoms in dialysis patients. *Am J Kidney Dis* 2017;**69**:117–28.
86. Hennessy MD, Zak RS, Gay CL, Pullinger CR, Lee KA, Aouizerat BE. Polymorphisms of interleukin-1 Beta and interleukin-17Alpha genes are associated with restless legs syndrome. *Biol Res Nurs* 2014;**16**:143–51.
87. Gan-Or Z, Zhou S, Ambalavanan A, Leblond CS, Xie P, Johnson A, et al. Analysis of functional GLO1 variants in the BTBD9 locus and restless legs syndrome. *Sleep Med* 2015;**16**:1151–5.

88. Fuh JL, Chung MY, Yao SC, Chen PK, Liao YC, Hsu CL, et al. Susceptible genes of restless legs syndrome in migraine. *Cephalalgia* 2016;**36**:1028–37.
89. Parish JM. Genetic and immunologic aspects of sleep and sleep disorders. *Chest* 2013;**143**:1489–99.
90. Szentkirályi A, Völzke H, Hoffmann W, Winkelmann J, Berger K. Lack of association between genetic risk loci for restless legs syndrome and multimorbidity. *Sleep* 2016;**39**:111–15.
91. Khan FH, Ahlberg CD, Chow CA, Shah DR, Koo BB. Iron, dopamine, genetics, and hormones in the pathophysiology of restless legs syndrome. *J Neurol* 2017;**264**:1634–41.
92. Cohen J. The birth of CRISPR Inc. *Science* 2017;**355**:680–4.
93. Cohen J. CRISPR patent ruling leaves license holders scrambling. *Science* 2017;**355**:786.
94. Jinek M, Chylinski K, Fonfara I, Fonfara I, Hauer M, Doudna JA, et al. A programmable dual-RNA-guided DNA endonuclease in adaptive bacterial immunity. *Science* 2012;**337**:816–21.
95. Shalem O, Sanjana NE, Zhang F. High-throughput functional genomics using CRISPR-Cas9. *Nat Rev Gen* 2015;**16**:299–311.
96. Lin J, Zhou Y, Liu J, Chen J, Chen W, Zhao S, et al. Progress and application of CRISPR/Cas technology in biological and biomedical investigation. *J Cell Biochem* 2017;**118**:3061–71.
97. Franco-Tormo MJ, Salas-Crisostomo M, Rocha NB, Budde H, Machado S, Murillo-Rodríguez E. CRISPR/Cas9, the powerful new genome-editing tool for putative therapeutics in obesity. *J Mol Neurosci* 2018;**65**:10–16.
98. Murillo-Rodríguez E, Barbosa Rocha N, Barciela Veras A, Budde H, Machado S. The end of snoring?: Application of CRISPR/Cas9 genome editing for sleep disorders. *Sleep Vigilance* 2018;**2**:13–21.
99. Toth LA, Bhargava P. Animal models of sleep disorders. *Comp Med* 2013;**63**:91–104.
100. Gillombardo CB, Darrah R, Dick TE. C57BL/6 J mouse apolipoprotein A2 gene is deterministic for apnea. *Respir Physiol Neurobiol* 2017;**235**:88–94.

Further Reading

Hobson JA. Dreaming as delirium: a mental status analysis of our nightly madness. *Semin Neurol* 1997;**17**:121–8.

Tafti M, Lammers GJ, Dauvilliers Y, Overeem S, Mayer G, Nowak J, et al. Narcolepsy-associated HLA Class I alleles implicate cell-mediated cytotoxicity. *Sleep* 2016;**39**:581–7.

CHAPTER

9

Alcoholism and Sleep

Rishi Sharma[1,2], Pradeep Sahota[1,2] and Mahesh M. Thakkar[1,2]

[1]Harry S. Truman Memorial Veterans Hospital, Columbia, MO, United States [2]Department of Neurology, University of Missouri, Columbia, MO, United States

OUTLINE

Molecular Sleep–Wake Cycle
DOI: https://doi.org/10.1016/B978-0-12-816430-3.00009-9

9.1 INTRODUCTION

Alcohol use disorder (AUD, also known as alcoholism) is a leading cause of preventable death and disability. It has a very high lifetime prevalence, affecting one third of the population, accounting for 3.5% of deaths every year.[1] There is ample evidence to suggest that sleep disruptions are associated with increased risk for substance abuse including alcoholism.[2] A primary reason for alcohol abuse is sleep abnormalities. People drink alcohol to treat their sleep problems. Estimates state that 6%–19% of the general population and 15%–28% of people with insomnia use alcohol to promote sleep (reviewed in Refs. 3, 4). In addition, subjective and objective indicators of sleep disruptions are considered as predictors of alcohol relapse in recovering alcoholics. It is estimated that, within the United States, the cost of alcohol-related problems exceeds $180 billion out of which more than $18 billion is associated with alcohol-related sleep disorders.[3] Although the effects of alcohol on sleep play a crucial role in alcohol abuse and its related problems, very little is known regarding how and where alcohol acts in the brain to cause sleep disruptions.

Understanding the causation for alcohol-induced sleep disruptions may be helpful in devising a strategy to prevent the development of AUD. However, the relationship between alcohol and sleep is complex and not yet thoroughly understood. As described in Fig. 9.1, alcohol is a potent somnogen and induces sleep by increasing the quality (shortening the latency to fall asleep and increasing delta activity during sleep)

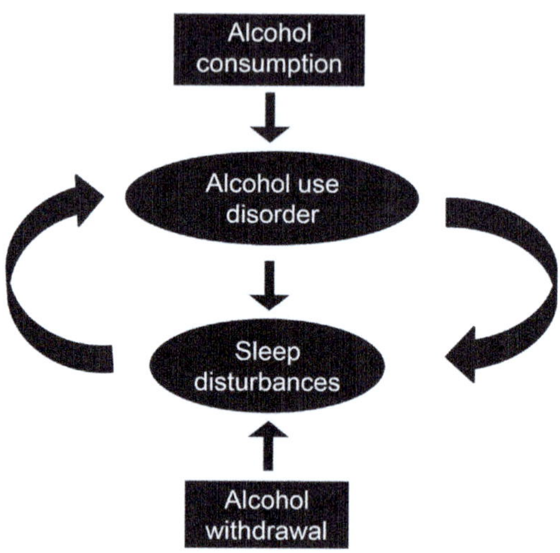

FIGURE 9.1 A complex interaction between alcohol and sleep.

and quantity (increasing duration) of sleep in healthy human volunteers. This sleep promoting activity makes alcohol one of the most common over the counter sleep aids.[5–7] Tolerance to these effects develops rapidly and such tolerance development may lead to excessive "hypnotic use" and possibly excessive daytime use of alcohol.[8] However, alcohol drinking is associated with severe sleep disruptions and the magnitude of severity increases with an increase in drinking.[3,9–11] Approximately 65% of heavy drinkers report serious insomnia and associated sleep problems.[10] The aim of this chapter is to discuss our previous findings regarding how and where alcohol acts to affect sleep. It begins with the description of the fundamentals of sleep regulation, which is followed by a review of the pertinent literature about the effects of alcohol on sleep in humans and animals. At the end, a summary of our published as well as unpublished/preliminary findings is described.

9.2 SLEEP–WAKEFULNESS

In this section, we will briefly describe the basic concepts of sleep–wakefulness, however, interested readers are referred to many in-depth reviews.[12–14] Sleep generally is defined as the behavior characterized by a reversible state of reduced voluntary movements along with the inability to respond to sensory stimulation. Sleep can be monitored by several objective and subjective methods, however, polysomnography (PSG) is considered the gold standard for assessing sleep physiology. Sleep monitoring via PSG involves monitoring of brain activity by electroencephalogram (or EEG), muscle movements by electromyogram (EMG), and eye movements by electrooculogram. Sleep promotion is the result of the gradual slowing of EEG with respect to high frequency and low-amplitude EEG signals, typical of wakefulness. Sleep comprises two distinct states: nonrapid eye movement (NREM) and rapid eye movement (REM) sleep.

In humans, sleep is monophasic and consists of alternate NREM and REM sleep cycles that take an average of 90 minutes and occurs four- to six times in a night sleep. While REM sleep is predominant in the second half, NREM sleep occupies the majority of time during the first half of sleep time and is divided into three stages. Stage 1 is the transition period between wakefulness and sleep, sometimes referred to as drowsiness; stage 2 is the first unequivocal stage of sleep during which muscle activity decreases further and conscious awareness of the outside world begins to fade; and stage 3 resembles the deepest sleep and shows characteristic high-amplitude delta waves (1–4 Hz) in the cortical EEG. Delta waves (also referred to as "delta activity" or "slow wave activity") are a measure of sleep intensity. Delta activity increases with sleep deprivation and is highest at the beginning of the sleep period. Naps during the day negatively affect delta activity in the subsequent night.[12]

In laboratory animals (mice, rats, and cats), the NREM sleep is generally not subdivided into stages. It is identified by the presence of high-amplitude, low-frequency waves in the EEG. In addition, the circadian distribution of the sleep period is polyphasic. The NREM–REM sleep cycles are shorter and occur throughout day and night during sleep.

During the passage from NREM to REM sleep, slow and synchronized cortical EEG signals transition to a fast, high frequency, low-amplitude activity, which appears similar to the desynchronized EEG of wakefulness. However, absence of muscle tone during REM sleep separates it from wakefulness. Besides muscle atonia, other important characteristics features of REM sleep are the presence of muscle twitches, ponto-geniculo-occipital waves, and bursts of REMs.[12,14]

9.2.1 Organization of Neuronal Network Regulating Sleep–Wakefulness

Multiple discrete centers or nuclei are present in the brain, which play an important role in the regulation of wakefulness. Classical work by Moruzzi and Magoun[15] has led to the identification of a set of ascending pathways, commonly termed as the ascending reticular activating system (ARAS), which originates from the brainstem and transmit information to the cortex. The ARAS is crucial for promoting wakefulness and is comprised of two relay pathways: dorsal and ventral. The dorsal pathway involves thalamic-relay nuclei, which receive excitatory input from cholinergic cell groups (acetylcholine-producing cells) in the upper brainstem: the pedunculopontine and laterodorsal tegmental nuclei (PPT/LDT) and transmits information to the cerebral cortex (Fig. 9.2[12,16]). The ventral pathway involves orexinergic neurons in the lateral hypothalamic area, which receive excitatory input from monoaminergic neurons in the upper brainstem and caudal hypothalamus, including the noradrenergic locus coeruleus, serotoninergic dorsal and median raphe nuclei, dopaminergic ventral periaqueductal gray matter, and histaminergic tuberomammillary neurons. The orexinergic hypothalamus sends projections to the basal forebrain, which is the terminal output nuclei of this pathway, and, in turn, sends excitatory projections to the cerebral cortex for activation.[13,17]

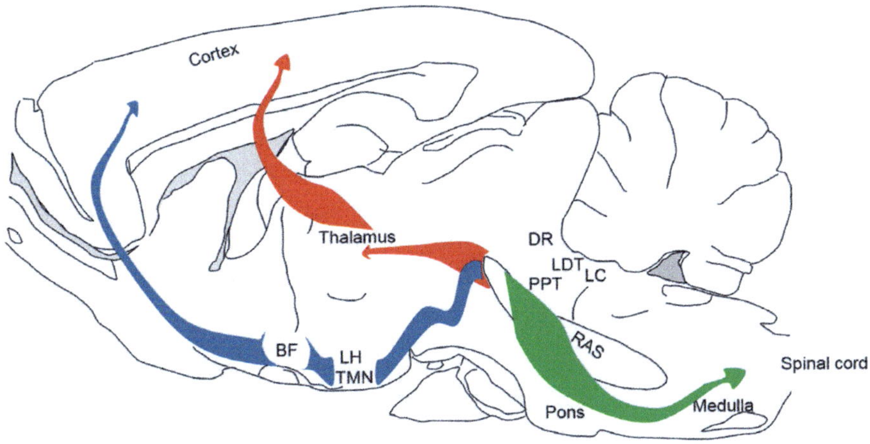

FIGURE 9.2 A schematic sagittal view of a mammalian brain depicting neuronal systems mediating the ventral (*blue*) and dorsal (*red*) relays of the ascending reticular activating system controlling cortical activation and behavioral arousal and neuronal systems (*green*) depicting descending projections to pons and medulla mediating muscle atonia during REM sleep. *BF*, cholinergic basal forebrain; *DR*, serotonergic dorsal raphe; *LC*, noradrenergic locus coeruleus; *LDT* and *PPT*, cholinergic laterodorsal and pedunculopontine tegmental nuclei; *LH*, orexinergic perifornical lateral hypothalamus; *RAS*, glutamatergic reticular activating system; *TMN*, histaminergic tuberomammillary nucleus.

The organizational network of REM sleep has two main components originating from cholinergic LDT/PPT nuclei. The first one includes the dorsal relay pathway, similar to that of wakefulness, which is responsible for the generation of fast and desynchronized cortical activity. The second one is responsible for muscle atonia. Overall, REM sleep is governed by the coordinated involvement of the pontine and medullary brainstem, which, via ascending projections, regulates the cortical desynchronization and, via descending projections, regulates muscle atonia (Fig. 9.2; reviewed in Refs. 12, 18).

The transition from wakefulness to NREM sleep involves a coordinated shutdown of the wake-promoting systems. Although there are multiple sleep active neurons such as GABA, and melanin concentrating hormone present in the median preoptic nucleus (MnPO), the ventrolateral preoptic area, and lateral and incerto-hypothalamic area,[12,19–22] adenosine, a homeostatic sleep factor, plays a major role in this process.[16]

9.2.1.1 Adenosine: A Sleep Homeostatic Factor

As discussed earlier, sleep homeostasis plays a crucial role in sleep initiation. The core of sleep homeostasis is sleep propensity, otherwise known as sleep pressure. Sleep pressure starts to accumulate as soon as one is awake and continues to accumulate until sleep is initiated. Once sleep is initiated, sleep pressure starts to dissipate. While sleep homeostasis is crucial for sleep induction, adenosine is crucial for the development of sleep pressure.[16] Experimentally, sleep homeostasis is examined by sleep deprivation followed by recovery sleep. During sleep deprivation, increase in sleep pressure is measured by increase in theta power in the wake EEG and accumulation of adenosine in the extracellular space, most markedly in the wake-promoting basal forebrain region. Increased adenosine acts via adenosine A1 receptor (A1R) to inhibit wake-promoting neurons in the basal forebrain to promote sleep and enhance delta activity.[23,24] Longer periods of wakefulness cause an accumulation of sleep pressure and/or adenosine, which results in a greater amount of time required for dissipation.[25]

9.2.2 Regulation of Sleep

Current theory suggests that the physiological regulation of mammalian sleep is controlled by two components: (1) a homeostatic sleep process (process S), which increases during wakefulness and declines during sleep and interacts with (2) a circadian alerting process (process C), that is independent of sleep and wakefulness.[26]

During normal spontaneous wakefulness, rising homeostatic sleep pressure is compensated for by an increase in the circadian alerting signal, which helps in the maintenance of wakefulness. Conversely, during the night, diminution of the circadian alerting signal coupled with residual homeostatic sleep drive ensures the maintenance of sleep.[27] The changes in sleep–wakefulness across 24 hours, observed in humans and animals, are the result of the combined influences of process S and process C (Fig. 9.3). The assessment of the relative contributions of sleep homeostasis and circadian rhythmicity occurs in an orchestrated fashion.[28]

Laboratory rodents are predominantly active during the dark period and sleep predominantly during the light period, alluding to process C. They display a shorter sleep cycle interspersed with awake episodes, mainly due to a limited capacity for accumulation of sleep pressures. Despite this "polyphasic" pattern of sleeping, these nocturnal laboratory rodents display a decrease in homeostatic drive during relatively consolidated periods of sleep and an increase during periods of wakefulness suggesting that process C and process S work in similar ways in humans and rodents.[29–31]

These fundamental elements of sleep are dynamic to outside influences, such as alcohol, which can alter sleep–wakefulness after varying amounts of consumption.

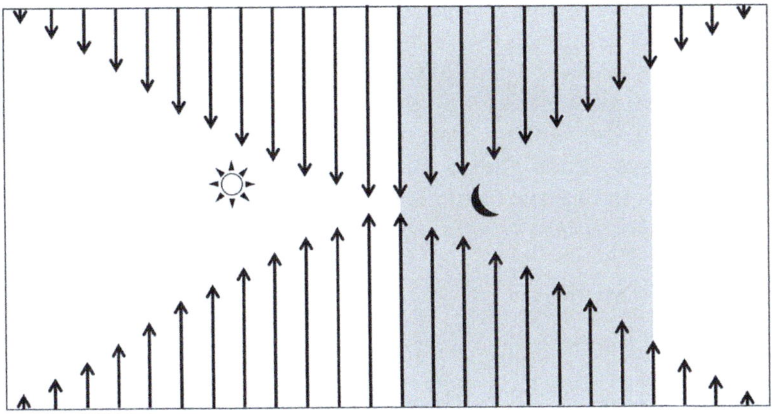

FIGURE 9.3 Schematic describing the interaction between the circadian alerting signal (process C) and the homeostatic sleep drive (process S) in the regulation of sleep–wakefulness. During the day, an increasing circadian alerting signal counteracts accumulating sleep pressure to maintain wakefulness. At night, the accumulating sleep pressure reaches its peak (threshold) whereby it overrides the circadian alerting drive and promotes sleep. During the night, sleep is maintained by the residual homeostatic sleep drive coupled with diminution of the circadian alerting signal (for details, see Ref. 27).

9.3 ALCOHOL AND SLEEP

Sleep plays an important role in alcohol abuse, particularly when disruptions in normal sleep pattern are involved. There is ample evidence to suggest that alcohol consumption is positively associated with trouble falling and staying asleep (reviewed in Ref. 10). According to Roehrs and colleagues, insomniacs are more likely to self-administer alcohol before bedtime than normal sleeping people.[32] In fact, sleep problems are a major contributing risk factor for alcohol abuse.[10,33–37] It is frequently asked why people who have sleep abnormalities drink alcohol as self-medication to intentionally induce sleep and what happens to their sleep as a result of this alcohol consumption. The following section discusses clinical and preclinical evidence to evaluate the effects of alcohol on sleep to propose an explanation for these questions.

9.3.1 Alcohol, its Dosages and Levels

In most studies, the amount of alcohol consumed is described as grams of pure ethanol/kilogram of the subject's body weight (expressed as g/kg). To determine ethanol content in the blood before performing a sleep test, studies have used either breath alcohol concentration (BrAC, expressed in grams per 100 mL or g%) or blood alcohol concentration (BAC, expressed in milligrams/100 mL or mg/dL or mg%) to provide a measure of the actual alcohol levels just prior to bedtime or sleep testing. In the United States, breath analyzers are calibrated by industry standards such that a BrAC of 0.10% is the equivalent of a BAC of 100 mg ethanol per 100 mL blood (mg/dL).

9.3.2 Clinical Studies

9.3.2.1 Acute Effects

A number of studies have been conducted to understand the effects of alcohol on sleep–wakefulness. Most studies, conducted in normal and healthy human subjects, have used high doses of alcohol (0.7–1 g/kg) achieving a BAC between 50 and 80 mg/dL. In general, these studies suggest that alcohol is a potent somnogen. Acute alcohol administered 30–60 minutes before bedtime in healthy nonalcoholics shortens sleep onset latency, and consolidates and increases the quality (delta power) and quantity of NREM sleep during the first half of the night,[16,38–44] hence providing a reasonable explanation as to why people with sleep problems use alcohol as a sleep aid.[5,10,33–37] However, interestingly, while generally alcohol at bedtime helps people to fall asleep and maintain sleep during the first few hours of the night, it has a very disruptive effect on the rest of the night, marked by fragmented sleep and frequent awakening.[5,8,9,39,40,42,45,46] This is generally viewed as a rebound effect because a BAC value of 80 mg/dL after alcohol consumption at bedtime is reduced to negligible value within 4–6 hours with a clearance rate of 10–20 mg/dL per hour.[41] Similarly, the effects on REM sleep in the first half of sleep appear to be dose related, that is, suppression of REM sleep in the first part of sleep with higher doses of alcohol at bedtime. In addition, a phenomenon called REM sleep rebound was observed by few studies,[39,42] but not by others.[47] All of the above studies focused on understanding the effects of alcohol on sleep when alcohol was consumed at bedtime, however, a majority of people consume alcohol during the popular "happy hour specials," usually offered between 5 and 7 p.m., 3–5 hours before bedtime.[48] What happens to their sleep? It has been shown that alcohol consumed even 6 hours before bedtime can increase wakefulness during the second half of sleep, even though the alcohol consumed has already been eliminated from the body by bedtime.[49] Overall, it is suggested that alcohol disrupts sleep, irrespective of timing.

Sleep disruptions are not the only negative consequence of alcohol consumption. Another more common negative consequence of alcohol abuse is having a "hangover," which is defined by the presence of several unpleasant/aversive symptoms including daytime sleepiness that are observed when BAC is 0.[50–52] The sleep disruptive effects of alcohol have been regarded as a major contributor to the hangover effects of alcohol on the following day.[52,53] Evidence suggests that individuals experiencing a hangover may consume more alcohol to alleviate the symptoms of their hangover commonly described by the expression "hair of the dog," resulting in the initiation of a vicious cycle that may eventually culminate in the development of AUD.[54,55]

While heavy drinking is a major risk factor for the development of AUD, tolerance to alcohol's aversive effects is considered as the first symptom of AUD. Tolerance develops over time with continuous use of alcohol. By definition, alcohol tolerance is the accentuated reduction in effect with the constant use of the same quantity or a need to progressively consume higher amounts of alcohol to achieve the same outcome.[56] Functional tolerance to alcohol is the result of the brain adapting to alcohol-induced disruption and is classified as (1) acute tolerance, observed during a single session of alcohol consumption; (2) rapid tolerance, observed between 8 and 24 hours after the effect of the first alcohol consumption has disappeared; and (3) chronic tolerance, which is detected after long-term (days) alcohol exposure.[57,58]

Previous studies suggest that tolerance develops from the NREM sleep promoting and REM sleep-reducing effects of alcohol during exposure to repeated episodes of alcohol consumption.[38,41] This is of major concern because tolerance to these effects develops rapidly and such tolerance development may lead to excessive "hypnotic use" and possibly excessive daytime use of alcohol.[8] There is ample evidence that suggests that development of rapid and acute tolerance to the behavioral impairing effects of alcohol is one of the major factors responsible for heavy drinking.[59,60]

9.3.2.2 *Chronic Effects*

Alcoholics experience problems falling asleep and decreased total sleep time during both drinking periods and withdrawal.[3] Several studies have investigated the effects of alcohol on sleep in alcoholics. To mimic heavy alcohol consumption in alcoholics, high dose of alcohol was used to achieve a BAC between 150 and 250 mg/dL.[61,62] Allen and colleagues have shown that administration of alcohol improves the sleep of alcoholics,[63] however, the majority of other studies suggest that alcoholics, following alcohol consumption, display reduced total time sleep, sleep fragmentation, an appearance of a lighter sleep stage 1, or wakefulness and an increased number of awakenings.[61,62,64] Taken together these findings suggest that in alcoholics, alcohol is more likely to aggravate sleep disturbances and may have minimal sedative effects.

To understand the effects of alcohol withdrawal on sleep–wakefulness, studies have investigated the sleep–wake behavior in alcoholics undergoing acute (abstinent for days[62] and weeks[65–67]) and chronic (abstinent for months[67–69]) alcohol withdrawal. In general, insomnia complaints are frequent and persist for weeks to months postabstinence.[70,71] Alcoholics, during recent abstinence (lasting for 2–8 weeks) display severe sleep problems as compared with nonalcoholics. The specific sleep–wake changes include decreased total sleep time, decreased sleep efficiency, and decreased NREM sleep along with an increase in sleep onset latency and stage 1 sleep.[72–75] Alcoholics normalize in total sleep time and sleep onset latency after 5–9 months, but still display sleep fragmentation during sustained abstinence (sober for 1–3 years).[3,68,75] The abnormalities in the REM sleep vary across studies. Drummond and colleagues[67] found an increase in REM sleep with a reduction in REM sleep latency while others found no change in REM sleep after 1 year of abstinence.[42,73] Relapse after alcohol abstinence is one of the biggest challenges in the treatment of alcoholism. Up to 85% of alcoholic patients relapse, independent of whether they have been treated as inpatients until complete remission of physical withdrawal symptoms.[76] Considering the long-term prevalence of sleep problems and their role in relapse in abstinent alcoholics, it is very important to understand the relationship between sleep problems and relapse to alcoholism.

There is ample evidence that suggests that sleep disruptions during alcohol withdrawal predict relapse (reviewed in Refs. 73, 77). Increased REM sleep and shortened REM sleep latency were associated with higher relapse rates during recent abstinence.[78,79] Of particular importance, untreated insomnia may interfere with recovery from the primary alcohol addiction and contribute to relapse within the first several months of recovery.[70] Thus, it is essential to understand the mechanism of insomnia associated with alcohol withdrawal to treat alcoholism.

9.3.3 Animal Studies

9.3.3.1 Acute Effects

Animal studies corroborate human studies. The doses of alcohol used in these studies were in the range from 2 to 3 g/kg achieving a BAC of >80 mg/dL, which is equivalent to the BAC achieved after 1.0 g of alcohol in humans. These studies suggest that irrespective of time of administration, acute alcohol treatment (either systemic or icv) in rodents produces a reduction in sleep onset latency and an increase in

NREM sleep, although REM sleep suppression was not consistently observed.[80–85] In addition, local bilateral infusion of alcohol into the preoptic region dose-dependently increases NREM sleep.[85]

9.3.3.2 *Chronic Studies*

A number of animal studies have been used to investigate the effects of chronic alcohol exposure on sleep–wakefulness using rats[80,86–91] and mice[92] as animal models. During the drinking period, rats display reversed sleep–wakefulness patterns during the light and dark periods, with more sleep during the dark (active) period and more wakefulness during the light (inactive) period.[80,87] Similar pattern was also observed during the withdrawal period.[86,87] The effects of alcohol withdrawal on REM sleep are inconclusive. While two studies have reported decreased REM sleep during withdrawal after chronic alcohol exposure,[89,93] two other studies reported increased REM sleep after withdrawal.[80,91] Veatch[92] found increased REM sleep with a concomitant reduction in NREM sleep during the initial withdrawal period. While REM sleep remained elevated, NREM sleep normalized within 4 days.

In summary, strong evidence from human and animal studies suggest that a single dose of alcohol reduces sleep onset latency and promotes NREM sleep. In humans, acute alcohol (especially a moderate to high dose) suppresses REM sleep primarily during the first half of the sleep period along with an increase in wakefulness during the second half. In contrast, both in humans and in animals, chronic alcohol exposure causes sleep disruptions, manifested by an increase in sleep onset latency and wakefulness along with a reduction in sleep during the normal sleep (light) period (insomnia), along with reduced wakefulness and increased sleep during the normal active period.

9.3.4 Anatomical and Cellular Mediators of Alcohol and Sleep

Considering the importance of alcohol-induced sleep problems in the development of alcoholism, several investigators have focused their research in identifying the target that could mediate the effects of alcohol on sleep. Multiple neuronal mediators that may mediate the effect of alcohol on sleep have been proposed such as serotonin, noradrenaline, glutamate, and GABA[16,94–101]; however, adenosine has gained a lot of attention. There are several reasons: adenosine may mediate the behavioral effects of alcohol and is implicated in sleep homeostasis.[24,102] Alcohol affects extracellular levels of adenosine. While acute exposure inhibits adenosine reuptake via an NBTI-sensitive equilibrative nucleoside transporter-1 (ENT1), chronic exposure downregulates ENT1 expression.[103,104] Genetic ablation of ENT1 in mice leads to a reduction in extracellular levels of adenosine and attenuated loss of righting reflex (hypnotic) and ataxic responses to alcohol along with increased alcohol consumption. In ENT1-null mice, administration of an A1R agonist decreases alcohol consumption whereas an A2AR agonist is ineffective in decreasing alcohol consumption.[105] There is ample evidence suggesting the role of adenosine and its receptors (A1R and A2aR) in mediating the neuronal and behavioral responses of alcohol, such as ataxia.[106–112] During alcohol withdrawal, adenosine is implicated in mediating behaviors such as anxiety, tremors, and seizures.[113–116] Several clinical studies also support the role of adenosine in alcohol-induced behaviors including motor performance, attention, and sleep.[117–126] However, the role of adenosine in mediating alcohol-induced changes in sleep–wakefulness has not been investigated.

To understand this, we employed multiple approaches using rats and mice as our animal models: (1) our first approach involved the use of acute administration (intragastric) of single high dose of alcohol in rats; (2) our second approach involved alcohol self-administration in mice for 4 hours; (3) in our third approach, we used a 1-day (episodic) binge protocol—to resemble the high level of alcohol intoxication usually achieved during binge drinking episodes, and that resembles high BAC—one that is in the range of those reported in trauma patients in emergency rooms across the United States (see Refs. 127–129); (4) in our fourth approach, we used a chronic mouse model of alcohol exposure[130–132]; and (5) in the fifth approach, we used Majchrowicz's rat model of alcohol dependence.

9.3.4.1 *Acute Administration*

Five experiments were performed on Sprague–Dawley (SD) rats. In all the experiments, alcohol (3 g/kg) was administered through intragastric gavage except in experiment 4, in which alcohol was directly infused in the basal forebrain using microdialysis.[84,133]

In the first experiment, we used c-Fos as a neuronal activation marker and examined the effects of alcohol on the activation of basal forebrain cholinergic wake-promoting neurons. We found that alcohol administration at dark onset significantly reduced the activation of wake-promoting neurons in the basal forebrain as compared with water administration (Fig. 9.4A).

Next, we examined the effectiveness of alcohol to promote sleep. Rats implanted with sleep recording electrodes (EEG and EMG) were administered sterile water (10 mL/kg) at dark onset on day 1 and alcohol (3 g/kg) on day 2 (a within-group design). We found that alcohol administration at dark onset significantly reduced sleep onset latency and increased NREM sleep (Fig. 9.4B). Wakefulness was significantly reduced but no change in REM sleep was observed. Temporally, the

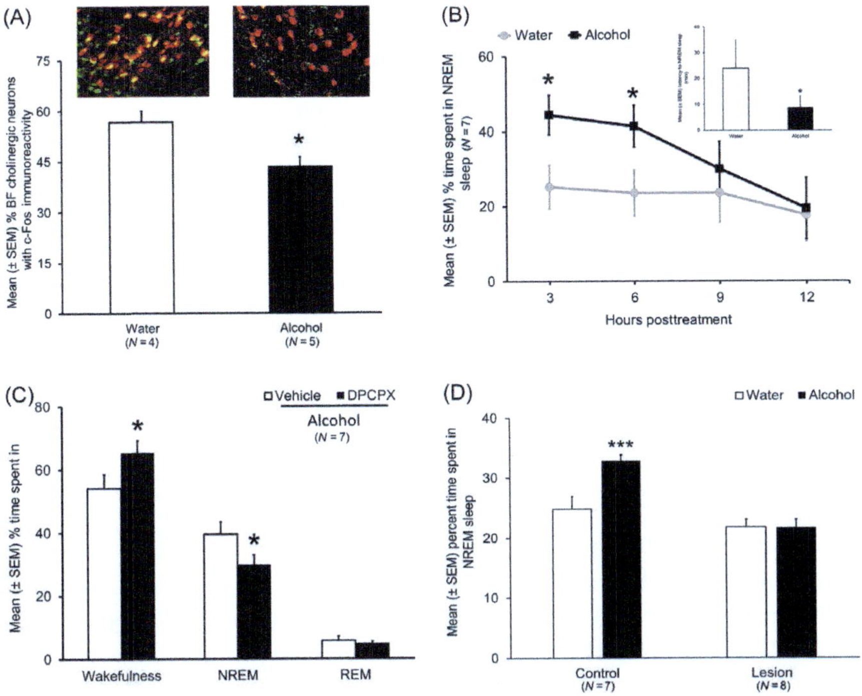

FIGURE 9.4 Acute alcohol administration interacts with the adenosinergic mechanism in the basal forebrain to promote sleep. A single dose (3 g/kg; intragastric) of alcohol, administered at the onset of the circadian active period, inhibits wake-promoting cholinergic neurons of the basal forebrain (A) and promotes NREM sleep as evident by a significant reduction in sleep onset latency (inset in B) coupled with a significant increase in the amount of time spent in NREM sleep (B). Blockade of adenosine A1R receptors specifically in the basal forebrain, by local microinjection of DPCPX, an A1R receptor antagonist, significantly attenuated alcohol-induced sleep promotion (C). Lastly, alcohol had a robust sleep promoting effect in controls (nonlesioned rats) as evident by a significant increase in NREM sleep. No such alcohol-induced sleep promotion was observed in lesioned rats (for details, see Ref. 134). $***P < .001$, $*P < .05$.

maximum effect of alcohol was observed during the first 6 h after administration. Our results suggest that alcohol is a potent somnogen and is able to counteract the strong circadian wake-promoting drive to promote NREM sleep.

Adenosine acts via two G-protein coupled receptors: A1R and A2AR. Since the basal forebrain lacks A2A receptors and adenosine promotes sleep by inhibiting the wake-promoting neurons of the basal forebrain via A1R,[23,24,119] we chose to examine the role of A1R in alcohol-induced sleep promotion. We used 8-cyclopentyl-1, 3-dipropylxanthine (DPCPX), which is an A1R receptor antagonist and promotes wakefulness.[135] It was found that bilateral infusion of DPCPX in the basal forebrain significantly attenuated alcohol-induced sleep promotion as compared with the vehicle (Fig. 9.4C) suggesting that adenosine mediates alcohol-induced sleep promotion.[84]

If the wake-promoting cholinergic neurons of basal forebrain are responsible for acute alcohol-induced sleep promotion, then lesions of the cholinergic neurons could attenuate the sleep promoting effects of alcohol. 192-IgG-saporin (SAP) was used to selectively lesion basal forebrain cholinergic neurons,[134] which resulted in a significant reduction (>70%) in the number of cholinergic neurons in the basal forebrain. Our results suggest that alcohol administration in lesioned animals resulted in an attenuated alcohol-induced sleep promotion (Fig. 9.4D).

Based on these findings, we suggest that alcohol promotes sleep by acting on and utilizing components of sleep homeostasis. It acts on the wake-promoting cholinergic neurons in the basal forebrain to increase extracellular levels of adenosine. Increased adenosine promotes sleep by inhibiting basal forebrain wake-promoting neurons via A1R.

9.3.4.2 *Self-Administration*

We have provided evidence to suggest that a single high dose of alcohol affects sleep homeostasis; however, all the above experiments involve the forced feeding of alcohol, which has several limitations. For example, gavage administration involves pain and stress in animals, which is likely to affect their natural sleep patterns. Moreover, humans self-administer alcohol. To mimic the conditions of human self-administration, we used a "drinking in the dark" (DID) procedure, which is relatively simple and frequently implemented to examine binge alcohol drinking in mice. This paradigm has high face validity in mimicking binge drinking in humans.[83,136–143] In this procedure, mice self-administer large amounts of alcohol within a relatively short period of time.[129]

The DID procedure was as follows: C57BL/6 J mice were allowed to voluntarily binge drink under a modified version of the DID paradigm by offering each mouse a single tube containing 20% alcohol (v/v; in tap water) instead of its usual water bottle, for a period of 4 hours, beginning 3 hours from the onset of the circadian dark period sleep.[139]

We performed five experiments using this DID procedure to understand the role of adenosinergic mechanisms in the basal forebrain in binge drinking–induced sleep disruptions.

We first examined the effects of binge alcohol consumption on sleep–wakefulness. In this experiment, mice implanted with sleep recording electrodes were exposed to water (on baseline day; baseline) and alcohol (on alcohol day; EtOH) using DID protocol. We recorded sleep–wakefulness for 21 hours and analyzed the results in three time bins: (1) during 4 hours of binge drinking (dark period), (2) during

4 hours of postbinge period (dark period), and (3) during 12 hours postbinge period (light period). We found that while sleep–wakefulness remained unchanged during the drinking period, a robust sleep promoting effect (>30% increase in NREM and REM sleep) was observed during the subsequent active (dark) period followed by severe sleep disturbance during the 12 hours of light/inactive period (Fig. 9.5A–C). Since the timing of sleep initiation at light onset was not affected, it is likely that circadian regulation of sleep remained unaffected.

To examine the effect of binge alcohol drinking on sleep homeostasis, the sleep deprivation recovery paradigm was used. Using this paradigm, the build-up of sleep-pressure, during 6 hours of sleep deprivation (highest at the end), can be examined by monitoring increase in the wake theta power. After sleep deprivation, dissipation of sleep pressure can be examined by indicators such as a reduction in latency to NREM sleep and increases in NREM delta power and the amount of time spent in NREM sleep during recovery sleep.[25,31,145–150]

Mice were divided in two groups: control and binge (EtOH). Mice in the control group were allowed to self-administer water whereas mice in the EtOH group were allowed to self-administer alcohol. In the following light period, sleep deprivation was performed in the last 6 hours (Fig. 9.5D). We found that during sleep deprivation, controls (exposed to water consumption) displayed an increase in wake theta power especially during the last 2 hours of sleep deprivation period, while no such effects were observed in mice exposed to alcohol suggesting no sleep pressure building up (Fig. 9.5E). Additionally, as compared with controls, mice in the alcohol group showed an increase in sleep onset latency, a reduction in the amount of time spent in the NREM sleep, and reduction in delta activity (Fig. 9.5F–H). These results suggest that a single episode of binge alcohol consumption disrupts sleep homeostasis.

Well-supported and convincing evidence exists to suggest that adenosine in the basal forebrain is a mediator of sleep homeostasis.[151] In the next experiment, we investigated the effects of binge drinking on extracellular levels of adenosine. Microdialysis was performed in the basal forebrain during 4 hours of DID procedure and during 6 hours of sleep deprivation (last 6 hours of light period) in the control and binge (EtOH) groups (Fig. 9.5I). Adenosine was measured in the microdialysate using the HPLC method. As described in Fig. 9.5J and K, during an alcohol drinking session, adenosine levels in the basal forebrain significantly increased during alcohol consumption. Furthermore, the controls showed an increase in basal forebrain adenosine levels during sleep deprivation, while mice exposed to alcohol did not show such an increase.

The above results suggest that mice exposed to binge alcohol drinking display blunted sleep homeostatic response. To further understand the mechanism by which alcohol impacts sleep homeostasis, we

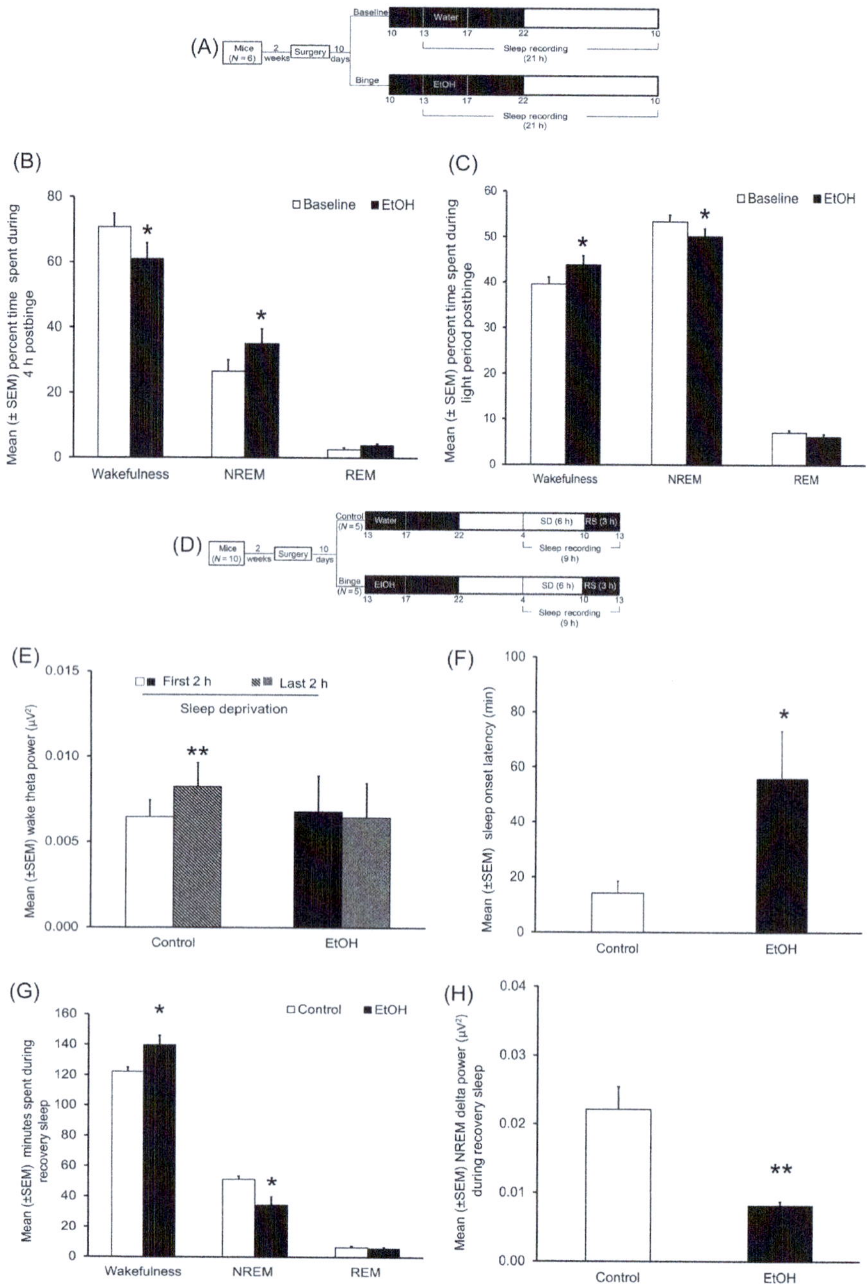

FIGURE 9.5 Acute self-administration of alcohol disrupts sleep. After 4 h of binge drinking (see Timeline in A), mice displayed significant increase in NREM sleep during dark period (B), however in the subsequent light period, insomnia was observed as evident by significant increase in percent time spent in wakefulness with a concomitant reduction in the NREM sleep (C). During sleep deprivation (D), theta activity was significantly increased in the control group but not in the EtOH group (E). During recovery sleep (D), the EtOH group showed a significant reduction in sleep onset latency at dark onset (F), a significant increase in recovery sleep (G) along with an increase in NREM delta activity (H) as compared with the controls. Microdialysis was performed during 4 h of binge

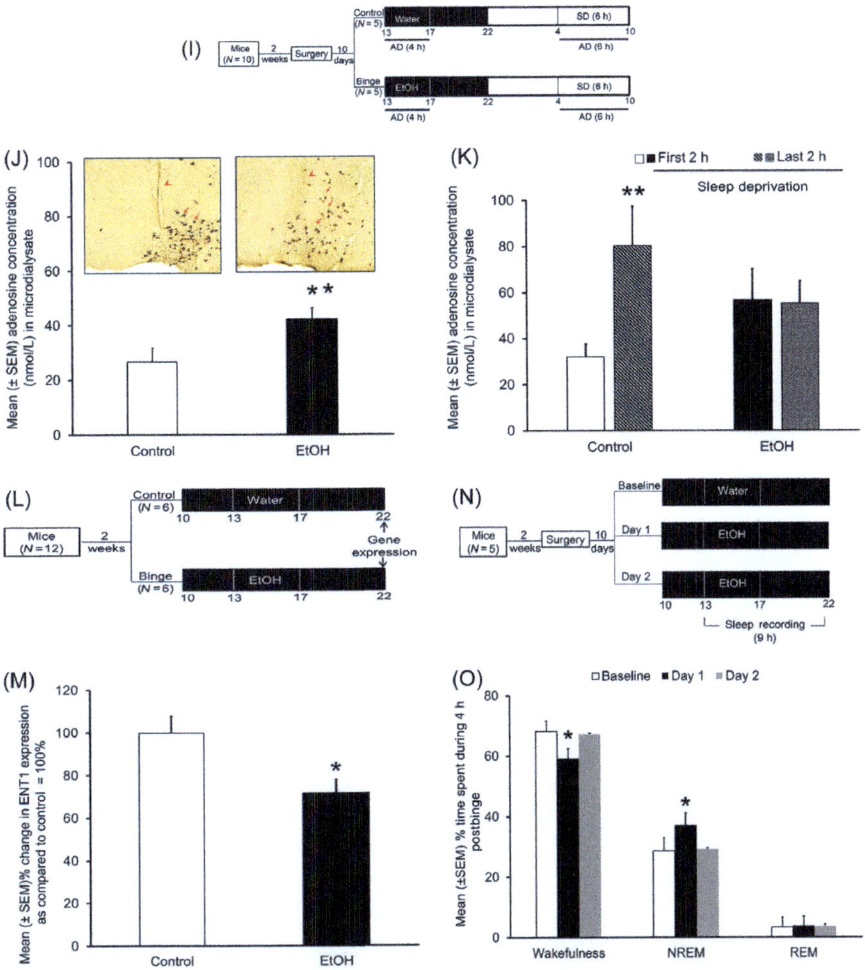

FIGURE 9.5 (Continued).

drinking and during 6 h of sleep deprivation (I). Although mice exposed to binge drinking displayed significant increase in the extracellular levels of adenosine in the basal forebrain during 4 h of alcohol consumption (J), they failed to display sleep deprivation-induced increase in adenosine release as seen in control mice (K). Gene expression of *ENT1* was examined at light onset to determine the underlying cause of altered sleep homeostatic mechanism (L). Mice exposed to binge alcohol drinking display a significant downregulation of *ENT1* gene expression in the basal forebrain as compared with controls (M). Binge drinking for two consecutive days at the same time (N) resulted in development of rapid tolerance as evident by a significant reduction of alcohol-induced sleep promotion on day 2 (O). All numbers below *black* and *white bars* describes military time. *Black bar* represents dark period and *white bar* represents light period. Drinking (water/alcohol) period (4 h) started at 13:00 (3 h after the dark onset) and ended at 17:00 in all the experiments (D). (For details, see Ref. 144). $^*P < .05$.

investigated the effects of alcohol self-administration on the gene expression of ENT1, ADK, and A1R, which are involved in the regulation of sleep homeostasis.[24,152,153] In this experiment (between-subject design), mice in the control and binge (EtOH) groups were euthanized at the onset of the light period (5 hours after the drinking stopped). This timeframe for euthanization was because binge alcohol drinking-induced sleep disruptions were observed during the light period (Fig. 9.5C). Using RT-PCR, we analyzed the gene expression of ENT1, ADK, and A1R. As shown in Fig. 9.5L and M, we found a significant reduction in ENT1 expression in the basal forebrain. The other two genes (A1R and ADK) remained unaffected. These results are in line with previous studies that implicate ENT1 in the regulation of adenosine release during spontaneous conditions, sleep deprivation, and after chronic alcohol exposure.[103,154,155] Interestingly, ENT1-null mice displayed a decreased adenosinergic tone, increased alcohol consumption, and reduced hypnotic (loss of righting reflex) response to alcohol.[105]

Since binge alcohol administration altered the sleep homeostatic mechanism, we hypothesized that subsequent exposure to binge alcohol exposure would display attenuated sleep promoting effects suggesting a development of rapid tolerance. To test this hypothesis, mice implanted with sleep recording electrodes were exposed to water on baseline day (baseline) and alcohol on day 1 and day 2 under the DID paradigm. Sleep–wakefulness data was analyzed for 3 days: baseline, day 1 and day 2. Mice consumed comparable amounts of alcohol on days 1 (Mean $\pm$ SEM = 3.67 $\pm$ 0.74 g/kg) and 2 (Mean $\pm$ SEM = 3.97 $\pm$ 0.67 g/kg). No change in sleep–wakefulness was observed during 4 hours of binge drinking on baseline, day 1, and day 2. However, during the 4 hours postbinge period, a robust alcohol-induced NREM sleep promotion was observed on day 1 as compared with baseline; however, no such sleep promotion was observed on day 2 suggesting a development of rapid tolerance (Fig. 9.5N and O) Based on our results, we suggest that altered sleep homeostatic mechanism following single binge exposure may be responsible for the development of rapid tolerance.

Overall, we found that binge alcohol-induced downregulation of ENT1 is associated with the altered adenosinergic mechanism in the basal forebrain resulting in sleep problems and tolerance to the sleep promoting effects of alcohol.

9.3.4.3 One Day Binge

Clinical evidence strongly suggests that sleep disruptions are one of the most common and severe consequences of alcohol intake with the magnitude of severity increasing as a result of binge drinking.[3,8–11] Approximately 65% of binge drinkers report serious insomnia and associated sleep problems.[10] Although the effects of alcohol on sleep play a crucial role in alcohol abuse and its related problems, very little is known regarding how binge alcohol consumption affects sleep regulation to cause sleep disruptions. Animal studies suggest that to mimic "binge" effects in animals, multiple alcohol exposures are required.[128] To follow these recommendations, we used a truncated version (1 day) of Majchrowicz's protocol as described previously.[154,156,157] Our protocol was as follows: rats were surgically implanted with sleep recording electrodes (as mentioned earlier) and allowed to recover from surgical stress and habituate with the sleep recording for 5–7 days. Once habituated, a baseline sleep–wake recording for 23 hours was performed. After baseline recording, binge alcohol treatment (three doses; 8 hours apart) was started 1 hour before light onset as described previously.[158] The doses of alcohol were adjusted based on the intoxication level.[84,154,157] The Mean ± SEM ($N = 7$) total dose of alcohol administered to rats is 14.6 ± 0.20 g/kg. The sleep–wakefulness recording was initiated at light onset on postbinge day 1 and continued until the end of the dark period on postbinge day 3. Major sleep–wakefulness changes were observed on postbinge day 2 after BAC was zero. We found that, as compared with baseline, the amount of time spent in wakefulness during the light (inactive) period was significantly increased (>40%), whereas the amount of time spent in NREM was significantly reduced (~17%), suggestive of insomnia. REM sleep remained unaffected (Fig. 9.6). Importantly, insomnia was observed only after alcohol was eliminated from the system (BAC = 0) suggesting

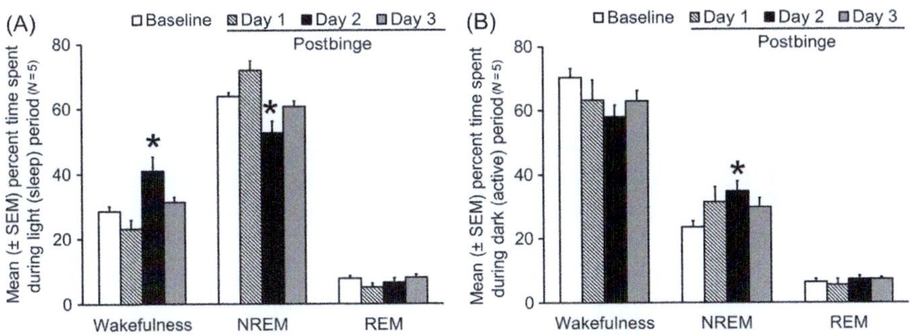

FIGURE 9.6 One-day binge drinking disrupts sleep–wakefulness. During the light (inactive) period, there was a significant increase in the amount of time spent in wakefulness with a concomitant reduction in the amount of time spent in NREM sleep on postbinge day 2 as compared with the baseline (A). REM sleep remained unaffected. No change in sleep–wakefulness on postbinge day 1 and the changes in sleep–wakefulness observed on postbinge day 2 were normalized on postbinge day 3. During the dark (active) period, significant increase in the amount of time spent in NREM sleep was observed during the dark period on postbinge day 2 (B). Wakefulness and REM sleep did not show any change. *$P < .05$. (For details, see Ref. 158).

that sleep disruptions accompany the "hangover/withdrawal" that is observed following a short-term binge (heavy) alcohol consumption.[50,51]

9.3.4.4 Chronic Alcohol Exposure

We used the NIAAA mice model of chronic alcohol exposure with slight modifications.[159] This model closely mimics clinical conditions of alcoholism with good face validity. For example, in this model, mice display (1) alcohol self-administration, (2) long-term heavy drinking, (3) development of tolerance, (4) escalation in alcohol intake with time, and (5) withdrawal symptoms.[130–132,160] This model has also been widely used to understand the brain mechanisms associated with alcohol withdrawal.[130–132,160–164] However, changes in sleep–wakefulness have never been examined in this chronic mouse model of alcohol consumption. In this between-group designed experiment, the animals were divided in two groups: alcohol and control. Animals in the alcohol group had ad lib access to the liquid diet (Lieber-Decarli '82 Shake and Pour control and ethanol liquid diet, Bio-Serv, Frenchtown, NJ) containing 6.8% (v/v) ethanol for 21 days. The control group received a specific amount of isocaloric control diet based on the average intake of the ethanol liquid diet in the alcohol group (~10 mL/24 hours) for the same duration. The procedure is as follows: for the first few days, the animals were acclimatized with the alcohol drinking procedure, which involved gradual increase in the concentration of ethanol in the liquid diet (2 days with control liquid diet, 2 days with 2.4%, and 2 days with 4.48% ethanol). Following this acclimatization period, animals could self-administer alcohol liquid diet containing 6.8% (v/v) ethanol for the next 21 days.

After 12 hours of alcohol exposure in the dark (on day 1), the mean ± SEM BAC was 124.7 ± 6.6 mg/dL ($N = 5$). Mice consumed significantly more alcohol on day 20 (24.3 ± 1.1 g/kg) as compared with day 2 (19.2 ± 1.5 g/kg). After 6 hours of alcohol withdrawal (on day 21), the BAC was undetectable suggesting that mice were in complete withdrawal.

Chronic alcohol exposure displays differential effects on the wakefulness across drinking days (Fig. 9.7). On day 2, there was a significant reduction in the wakefulness with a concomitant increase in the NREM sleep suggestive of the sleep-promoting effects of alcohol in the beginning of chronic alcohol exposure whereas, on day 20, a significant increase in the wakefulness was found along with a reduction in the NREM sleep suggestive of insomnia in mice (Fig. 9.7A–C).

During withdrawal, mice in the alcohol group displayed significant increase in wakefulness with a concomitant reduction in NREM sleep on withdrawal days 1, 3, and 5 as compared with the controls. The

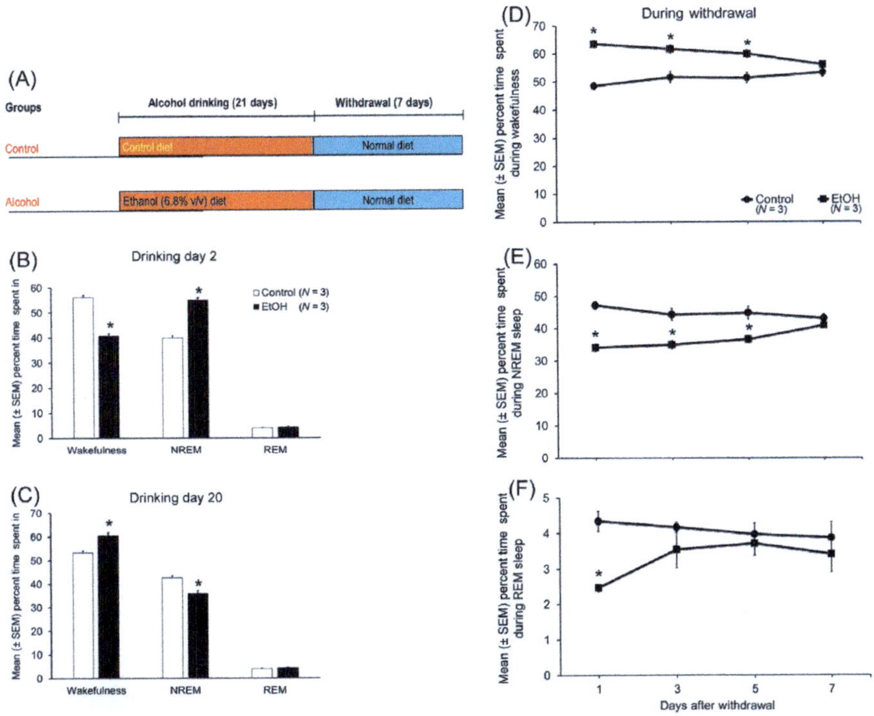

FIGURE 9.7 Chronic self-administration of alcohol causes insomnia. The timeline is described in (A). During day 2 of active drinking, mice in the alcohol group displayed significant reduction in the wakefulness with a concomitant increase in the NREM sleep, however, a nonsignificant increase in the REM was observed (B). In contrast, insomnia-like symptoms were observed on day 20 of active drinking as evident by a significant increase in the wakefulness with a concomitant reduction in NREM sleep as compared with controls (C). REM sleep remained unchanged (C). During withdrawal, insomnia was observed in the alcohol group as evident by significant increase in the wakefulness (D) with a concomitant reduction in the NREM sleep (E) on withdrawal days 1, 3, and 5 as compared with the controls. REM sleep was significantly reduced on withdrawal day 1 (F). $^*P < .05$.

REM sleep was significantly reduced on withdrawal day 1. Sleep–wakefulness became normalized by day 7 (Fig. 9.7D–F).

Evidence from clinical studies suggests that continuous long-term binge drinking is associated with the development of alcohol dependence or alcoholism.[165–167] To understand the mechanism of sleep problems associated with alcohol dependence or alcohol withdrawal, we used the rat model of alcohol dependence in SD rats as discussed below.

9.3.4.5 Alcohol Dependence

We used Majchrowicz's method to induce alcohol dependence/withdrawal in SD rats.[157] The procedure is similar as described in the 1-day binge protocol except alcohol was administered for 4 days (alcohol dependent group). Withdrawal day began at light onset, 10 hours after the last dose of alcohol. Sleep–wakefulness was examined during alcohol withdrawal.

As shown in Fig. 9.8A, during acute alcohol withdrawal, rats displayed significant increases in wakefulness, both during their sleep (light) and active (dark) period with a concomitant reduction in sleep (NREM and REM). This is suggestive of severe insomnia during alcohol withdrawal and supports clinical observations.[16,154]

In addition, chronic binge-exposed rats showed an increased activity of basal forebrain wake-promoting neurons during withdrawal as evident by a significant increase in the number of cholinergic neurons with c-Fos expression in the basal forebrain (Fig. 9.8B). This result implicates the role of cholinergic neurons in withdrawal-induced insomnia.

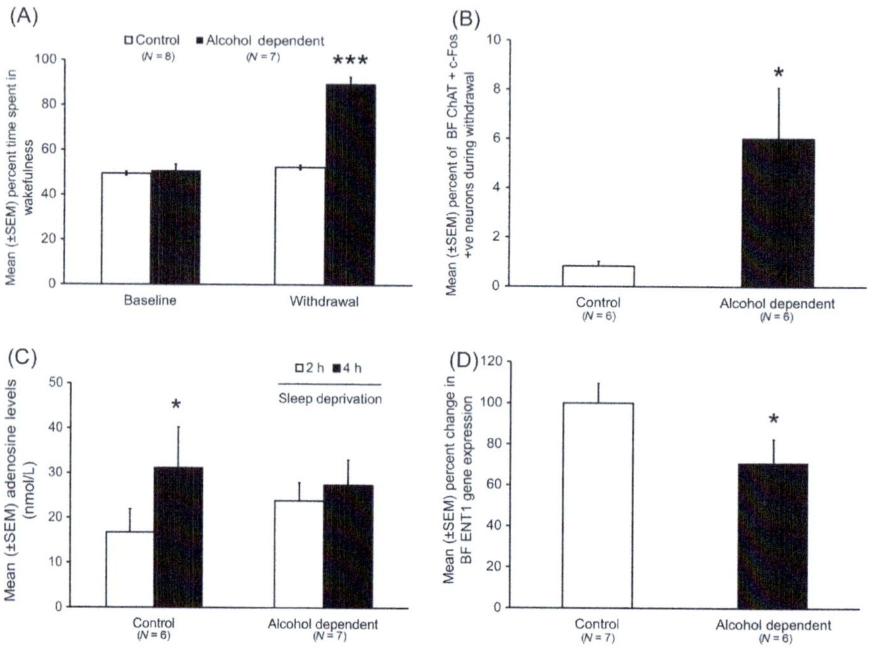

FIGURE 9.8 Alcohol dependence alters sleep homeostasis resulting in insomnia. As compared with controls, alcohol dependent rats displayed a significant increase in the amount of time spent in wakefulness (24 h; A) along with an increased activation of wake-promoting cholinergic neurons of the basal forebrain (B) during withdrawal. In addition, during withdrawal, alcohol dependent rats display an impairment in sleep deprivation-induced increase in adenosine in the basal forebrain (C), which may be due to a significant reduction in the expression of *ENT1* (D; for details see Ref. 154). $*P < .05$.

Furthermore, rats undergoing alcohol withdrawal showed impaired response to sleep deprivation as evident by inability to increase in extracellular adenosine during sleep deprivation suggestive of an impaired sleep homeostatic response (Fig. 9.8C).

Alcohol-dependent rats showed a reduction in the expression of *ENT1* genes in the basal forebrain, which may be responsible for altered sleep homeostasis (Fig. 9.8D). These results provided strong support for the role of adenosine in alcohol's effects on sleep.[16,154] In support of this claim, recent studies have confirmed that human alcoholics display a blunted sleep homeostatic response.[168,169]

There is ample evidence that strongly suggests that epigenetic mechanisms play a critical role in the development of addictive behavior, especially alcohol-related disorders.[170–174] Epigenetic mechanisms, such as DNA methylation and histone modifications (acetylation, methylation, and phosphorylation) are involved in the regulation of gene transcription. Previously, we found that alcohol dependent rats display downregulation of genes in the basal forebrain involved in sleep homeostasis (see Fig. 9.9D). Since histone acetylation (AcHis) and CREB-binding protein (CBP) increase gene transcription, we hypothesized that rats with alcohol dependence will display reduced expression of AcHis and CBP. Therefore, we performed IH for both AcHis and CBP in the basal forebrain. We found that there was a significant reduction in the expression of AcHis and CBP in the basal forebrain of alcohol dependent rats (Fig. 9.9A–F) suggesting that an altered epigenetic mechanism may be responsible for downregulation of genes involved in sleep homeostasis.

Trichostatin A (TSA; a histone deacetylases inhibitor) has recently been shown to reverse alcohol-induced changes in gene expression, ameliorate alcohol-induced behaviors such as anxiety, and also reduce alcohol consumption.[176–178] Hence, we hypothesized that TSA treatment during chronic alcohol exposure may reduce insomnia during alcohol withdrawal. To test our hypothesis, we use electrographic measures to examine the effects of TSA treatment on insomnia during withdrawal. Mice were divided in two groups: DMSO and TSA. Both groups were allowed to self-administer a liquid diet containing 6.8% v/v ethanol (as described above). However, mice in the TSA group received intraperitoneal injections of TSA (2 mg/kg; TSA group) at light onset on alternative days starting from day 15 of the alcohol drinking procedure till withdrawal day 1 (total of five doses). Mice in the DMSO group (controls) received 10% DMSO instead of TSA at the same time points. Our preliminary results suggest that mice ($N = 3$) treated with TSA displayed a significant reduction in the wakefulness and an increase in NREM sleep on day 1 as compared with the DMSO treated animals. Also, REM sleep was also significantly increased on days 1 and 2 of the

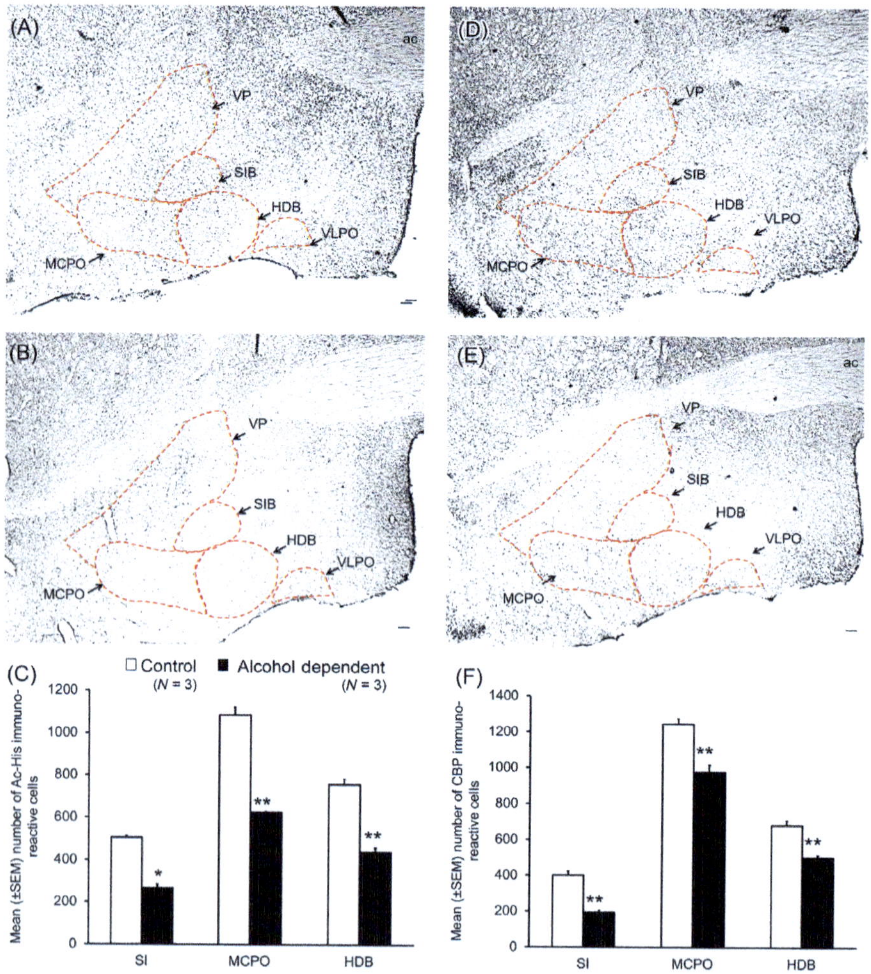

FIGURE 9.9 Alcohol dependence alters epigenetic mechanisms in the basal forebrain. Representative photomicrographs showing acetylated histone (AcHis; A and B) and CREB-binding protein (CBP; D and E) staining in the basal forebrain of controls (A and D) and alcohol dependent rats (B and E). Alcohol dependent rats displayed a significant reduction in the expression of AcHis (C) and CBP (F) in the basal forebrain as compared with the controls. $^{*}P < .05$, $^{**}P < .01$. *ac*, anterior commissure; *HDB*, horizontal diagonal band; *MCPO*, magnocellularis preoptic; *SIB*, substantia innominata. For identification of unlabeled structures see Ref. 175.

withdrawal (Fig. 9.10A–D). Although we did not investigate the mechanism or identify the target gene responsible for the effects of TSA on sleep disruptions, based on our previous findings (see Fig. 9.8)[154] we believe that TSA administration may ameliorate the expression of genes implicated in the regulation of sleep homeostasis.

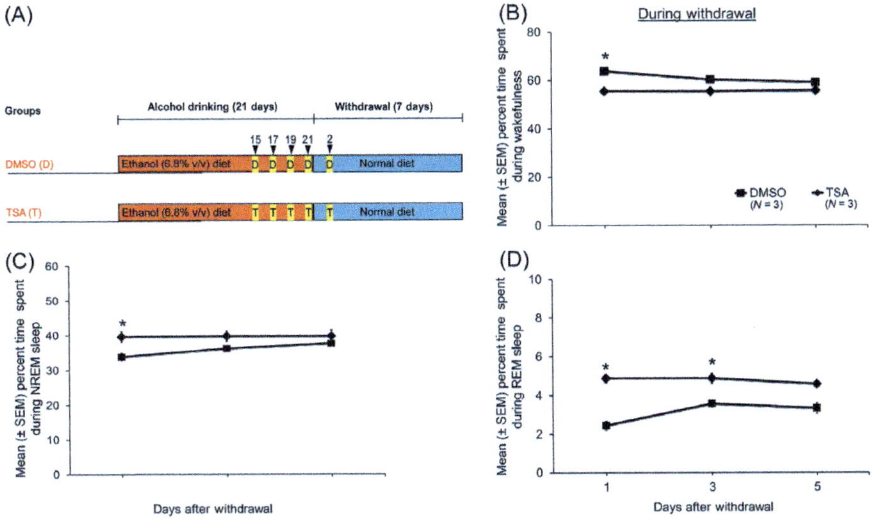

FIGURE 9.10 Epigenetic mechanism is involved in chronic alcohol-induced sleep disruptions. The timeline is described (A). Systemic administration of trichostatin A (TSA, a histone deacetylase inhibitor) in mice exposed to chronic alcohol significantly attenuated the symptoms of insomnia as evident by a significant reduction in the wakefulness (B) along with an increase in NREM sleep (C) during 24 h of day 1 as compared with the DMSO treated animals. Also, there was a significant increase in the REM sleep on days 1 and 3 (D). $*P < .05$. *D*, DMSO; *T*, TSA.

9.4 SUMMARY

Based on our results from acute and chronic studies, we suggest that alcohol's effects on sleep are mediated by its interaction with the sleep homeostatic mechanism. While alcohol-induced activation of the sleep homeostatic mechanism is responsible for its sleep promoting effects, the disruption in the sleep homeostatic mechanism caused by alcohol result in sleep problems. Moreover, disruption in the sleep homeostatic mechanism may also be responsible for the development of rapid tolerance to the sleep promoting effects of alcohol. Furthermore, our results also implicate the role of epigenetic mechanisms in alcohol-induced altered sleep homeostatic mechanism resulting in sleep disruptions.

Acknowledgments

This work was supported by the Harry S. Truman Memorial Veterans Hospital, and the funding from the National Institute of Alcohol Abuse and Alcoholism. We would also like to acknowledge the help of Dylan Clemmons, Destiny Heyer, and Abhilasha Sharma in the preparation of this manuscript.

References

1. Grant BF, Goldstein RB, Saha TD, Chou SP, Jung J, Zhang H, et al. Epidemiology of DSM-5 alcohol use disorder: results from the national epidemiologic survey on alcohol and related conditions III. *JAMA Psychiatry* 2015;**72**(8):757–66.
2. Shibley HL, Malcolm RJ, Veatch LM. Adolescents with insomnia and substance abuse: consequences and comorbidities. *J Psychiatry Pract* 2008;**14**(3):146–53.
3. Brower KJ. Alcohol's effects on sleep in alcoholics. *Alcohol Res Health* 2001; **25**(2):110–25.
4. Brower KJ. Assessment and treatment of insomnia in adult patients with alcohol use disorders. *Alcohol* 2015;**49**(4):417–27.
5. Johnson EO, Roehrs T, Roth T, Breslau N. Epidemiology of alcohol and medication as aids to sleep in early adulthood. *Sleep* 1998;**21**(2):178–86.
6. Goodhines PA, Gellis LA, Kim J, Fucito LM, Park A. Self-medication for sleep in college students: concurrent and prospective associations with sleep and alcohol behavior. *Behav Sleep Med* 2017;1–15.
7. Daley M, Morin CM, LeBlanc M, Grégoire JP, Savard J. The economic burden of insomnia: direct and indirect costs for individuals with insomnia syndrome, insomnia symptoms, and good sleepers. *Sleep* 2009;**32**(1):55–64.
8. Roehrs T, Roth T. Sleep, sleepiness, sleep disorders and alcohol use and abuse. *Sleep Med Rev* 2001;**5**(4):287–97.
9. Roehrs T, Roth T. Sleep, sleepiness, and alcohol use. *Alcohol Res Health* 2001; **25**(2):101–9.
10. Popovici I, French MT. Binge drinking and sleep problems among young adults. *Drug Alcohol Depend* 2013;**132**:207–15.
11. Vinson DC, Manning BK, Galliher JM, Dickinson LM, Pace WD, Turner BJ. Alcohol and sleep problems in primary care patients: a report from the AAFP National Research Network. *Ann Fam Med* 2010;**8**(6):484–92.
12. Brown RE, Basheer R, McKenna JT, Strecker RE, McCarley RW. Control of sleep and wakefulness. *Physiol Rev* 2012;**92**(3):1087–187.
13. McCarley RW. Neurobiology of REM and NREM sleep. *Sleep Med* 2007;**8**(4):302–30.
14. Steriade M, McCarley RW. *Brain control of sleep and wakefulness*. New York: Kluwer-Elsevier; 2005.
15. Moruzzi G, Magoun HW. Brain stem reticular formation and activation of the EEG. *Electroencephalogr Clin Neurophysiol* 1949;**1**:455–73.
16. Thakkar MM, Sharma R, Sahota P. Alcohol disrupts sleep homeostasis. *Alcohol* 2015; **49**(4):299–310.
17. Thakkar MM, Delgiacco RA, Strecker RE, McCarley RW. Adenosinergic A1 inhibition of basal forebrain wake-active neurons: a combined unit recording and microdialysis study in freely behaving cats. *Sleep* 1999;**22**:9.
18. Arrigoni E, Chen MC, Fuller PM. The anatomical, cellular and synaptic basis of motor atonia during rapid eye movement sleep. *J Physiol* 2016;**594**(19):5391–414.
19. Uschakov A, Gong H, McGinty D, Szymusiak R. Efferent projections from the median preoptic nucleus to sleep- and arousal-regulatory nuclei in the rat brain. *Neuroscience* 2007;**150**(1):104–20.
20. Jego S, Glasgow SD, Herrera CG, Ekstrand M, Reed SJ, Boyce R, et al. Optogenetic identification of a rapid-eye-movement (REM) sleep modulatory circuit in the hypothalamus. *Nat Neurosci* 2013;**16**(11):1637–43.
21. Vetrivelan R, Kong D, Ferrari LL, Arrigoni E, Madara JC, Bandaru SS, et al. Melanin-concentrating hormone neurons specifically promote rapid eye movement sleep in mice. *Neuroscience* 2016;**336**:102–13.

22. Konadhode RR, Pelluru D, Blanco-Centurion C, Zayachkivsky A, Liu M, Uhde T, et al. Optogenetic stimulation of MCH neurons increases sleep. *J Neurosci* 2013; **33**(25):10257–63.
23. Thakkar MM, Delgiacco RA, Strecker RE, McCarley RW. Adenosinergic inhibition of basal forebrain wakefulness—active neurons: a simultaneous unit recording and microdialysis study in freely behaving cats. *Neuroscience* 2003;**22**(4):1107–13.
24. Thakkar MM, Winston S, McCarley RW. A1 receptor and adenosinergic homeostatic regulation of sleep-wakefulness: effects of antisense to the A1 receptor in the cholinergic Basal forebrain. *J Neurosci* 2003;**23**(10):4278–87.
25. Porkka-Heiskanen T. Sleep homeostasis. *Curr Opin Neurobiol* 2013;**23**(5):799–805.
26. Borbely AA. A two process model of sleep regulation. *Human Neurobiol* 1982;**1**:195–204.
27. Edgar DM, Dement WC, Fuller CA. Effect of SCN lesions on sleep in squirrel monkeys: evidence for opponent processes in sleep-wake regulation. *J Neurosci* 1993; **13**(3):1065–79.
28. Tobler I, Borbely AA, Groos G. The effect of sleep deprivation on sleep in rats with suprachiasmatic lesions. *Neurosci Lett* 1983;**42**(1):49–54.
29. Franken P, Dijk DJ, Tobler I, Borbely AA. Sleep deprivation in rats: effects on EEG power spectra, vigilance states, and cortical temperature. *Am J Physiol* 1991;**261**(1 Pt 2): R198–208.
30. Tobler I, Franken P, Trachsel L, Borbely AA. Models of sleep regulation in mammals. *J Sleep Res* 1992;**1**(2):125–7.
31. Vyazovskiy VV, Tobler I. Theta activity in the waking EEG is a marker of sleep propensity in the rat. *Brain Res* 2005;**1050**(1–2):64–71.
32. Roehrs T, Papineau K, Rosenthal L, Roth T. Ethanol as a hypnotic in insomniacs: self administration and effects on sleep and mood. *Neuropsychopharmacology* 1999; **20**(3):279–86.
33. Wong M, Brower K, Fitzgerald H, Zucker R. Sleep problems in early childhood and early onset of alcohol and other drug use in adolescence. *Alcohol Clin Exp Res* 2004; **28**(4):578–87.
34. Hasler BP, Kirisci L, Clark DB. Restless sleep and variable sleep timing during late childhood accelerate the onset of alcohol and other drug involvement. *J Stud Alcohol Drugs* 2016;**77**(4):649–55.
35. Wong MM, Robertson GC, Dyson RB. Prospective relationship between poor sleep and substance-related problems in a national sample of adolescents. *Alcohol Clin Exp Res* 2015;**39**(2):355–62.
36. Canham SL, Kaufmann CN, Mauro PM, Mojtabai R, Spira AP. Binge drinking and insomnia in middle-aged and older adults: the Health and Retirement Study. *Int J Geriatr Psychiatry* 2015;**30**(3):284–91.
37. Haario P, Rahkonen O, Laaksonen M, Lahelma E, Lallukka T. Bidirectional associations between insomnia symptoms and unhealthy behaviours. *J Sleep Res* 2013;**22** (1):89–95.
38. Yules RB, Freedman DX, Chandler KA. The effect of ethyl alcohol on man's electroencephalographic sleep cycle. *Electroencephalogr Clin Neurophysiol* 1966;**20**(2):109–11.
39. Rundell OH, Lester BK, Griffiths WJ, Williams HL. Alcohol and sleep in young adults. *Psychopharmacologia* 1972;**26**(3):201–18.
40. Maclean AW, Cairns J. Dose-response effects of ethanol on the sleep of young men. *J Stud Alcohol* 1982;**43**(5):434–44.
41. Prinz PN, Roehrs TA, Vitaliano PP, Linnoila M, Weitzman ED. Effect of alcohol on sleep and nighttime plasma growth hormone and cortisol concentrations. *J Clin Endocrinol Metab* 1980;**51**(4):759–64.

42. Williams DL, Maclean AW, Cairns J. Dose-response effects of ethanol on the sleep of young women. *J Stud Alcohol* 1983;**44**(3):515–23.
43. Gresham SC, Webb WB, Williams RL. Alcohol and caffeine: effect on inferred visual dreaming. *Science* 1963;**140**:1226–7.
44. Zarcone Jr VP, Schreier L, Mitchell G, Orenberg E, Barchas J. Sleep variables, cyclic AMP and biogenic amine metabolites after one day of ethanol ingestion. *J Stud Alcohol* 1980;**41**(3):318–24.
45. Feige B, Gann H, Brueck R, Hornyak M, Litsch S, Hohagen F, et al. Effects of alcohol on polysomnographically recorded sleep in healthy subjects. *Alcohol Clin Exp Res* 2006;**30**(9):1527–37.
46. Arnedt JT, Rohsenow DJ, Almeida AB, Hunt SK, Gokhale M, Gottlieb DJ, et al. Sleep following alcohol intoxication in healthy, young adults: effects of sex and family history of alcoholism. *Alcohol Clin Exp Res* 2011;**35**(5):870–8.
47. Ebrahim IO, Shapiro CM, Williams AJ, Fenwick PB. Alcohol and sleep I: effects on normal sleep. *Alcohol Clin Exp Res* 2013;**37**(4):539–49.
48. Baldwin JM, Stogner JM, Miller BL. It's five o'clock somewhere: an examination of the association between happy hour drinking and negative consequences. *Subst Abuse Treat Prev Policy* 2014;**9**:17.
49. Landolt HP, Roth C, Dijk DJ, Borbely AA. Late-afternoon ethanol intake affects nocturnal sleep and the sleep EEG in middle-aged men. *J Clin Psychopharmacol* 1996;**16**(6):428–36.
50. Penning R, van NM, Fliervoet LA, Olivier B, Verster JC. The pathology of alcohol hangover. *Curr Drug Abuse Rev* 2010;**3**(2):68–75.
51. Verster JC, Stephens R, Penning R, Rohsenow D, McGeary J, Levy D, et al. The alcohol hangover research group consensus statement on best practice in alcohol hangover research. *Curr Drug Abuse Rev* 2010;**3**(2):116–26.
52. Prat G, Adan A, Sanchez-Turet M. Alcohol hangover: a critical review of explanatory factors. *Hum Psychopharmacol* 2009;**24**(4):259–67.
53. McKinney A, Coyle K. Alcohol hangover effects on measures of affect the morning after a normal night's drinking. *Alcohol Alcohol* 2006;**41**(1):54–60.
54. Piasecki TM, Robertson BM, Epler AJ. Hangover and risk for alcohol use disorders: existing evidence and potential mechanisms. *Curr Drug Abuse Rev* 2010;**3**(2):92–102.
55. Epler AJ, Tomko RL, Piasecki TM, Wood PK, Sher KJ, Shiffman S, et al. Does hangover influence the time to next drink? An investigation using ecological momentary assessment. *Alcohol Clin Exp Res* 2014;**38**(5):1461–9.
56. Tabakoff B, Cornell N, Hoffman PL. Alcohol tolerance. *Ann Emerg Med* 1986;**15**(9):1005–12.
57. Khanna JM, Kalant H, Shah G, Weiner J. Rapid tolerance as an index of chronic tolerance. *Pharmacol Biochem Behav* 1991;**38**(2):427–32.
58. Schuckit MA. Biological markers in alcoholism. *Prog Neuro-Psychopharmacol Biol Psychiatry* 1986;**10**(2):191–9.
59. Marczinski CA, Combs SW, Fillmore MT. Increased sensitivity to the disinhibiting effects of alcohol in binge drinkers. *Psychol Addict Behav* 2007;**21**(3):346–54.
60. Fillmore MT, Weafer J. Acute tolerance to alcohol in at-risk binge drinkers. *Psychol Addict Behav* 2012;**26**(4):693–702.
61. Mello NK, Mendelson JH. Behavioral studies of sleep patterns in alcoholics during intoxication and withdrawal. *J Pharmacol Exp Ther* 1970;**175**(1):94–112.
62. Johnson LC, Burdick JA, Smith J. Sleep during alcohol intake and withdrawal in the chronic alcoholic. *Arch Gen Psychiatry* 1970;**22**(5):406–18.
63. Allen RP, Wagman AM, Funderburk FR, Wells DT. Slow wave sleep: a predictor of individual differences in response to drinking? *Biol Psychiatry* 1980;**15**(2):345–8.

64. Gross MM, Goodenough DR, Hastey J, Lewis E. Experimental study of sleep in chronic alcoholics before, during, and after four days of heavy drinking with a non-drinking comparison. *Ann N Y Acad Sci* 1973;**215**:254–65.
65. Snyder S, Karacan I. Sleep patterns of sober chronic alcoholics. *Neuropsychobiology* 1985;**13**(1–2):97–100.
66. Irwin M, Miller C, Gillin JC, DeModena A, Ehlers CL. Polysomnographic and spectral sleep EEG in primary alcoholics: an interaction between alcohol dependence and African-American ethnicity. *Alcohol Clin Exp Res* 2000;**24**(9):1376–84.
67. Drummond SP, Gillin JC, Smith TL, DeModena A. The sleep of abstinent pure primary alcoholic patients: natural course and relationship to relapse. *Alcohol Clin Exp Res* 1998;**22**(8):1796–802.
68. Adamson J, Burdick JA. Sleep of dry alcoholics. *Arch Gen Psychiatry* 1973;**28**(1):146–9.
69. Colrain IM, Turlington S, Baker FC. Impact of alcoholism on sleep architecture and EEG power spectra in men and women. *Sleep* 2009;**32**(10):1341–52.
70. Arnedt JT, Conroy DA, Brower KJ. Treatment options for sleep disturbances during alcohol recovery. *J Addict Dis* 2007;**26**(4):41–54.
71. Benca RM, Obermeyer WH, Thisted RA, Gillin JC. Sleep and psychiatric disorders. A meta-analysis. *Arch Gen Psychiatry* 1992;**49**(8):651–68 discussion 69–70.
72. Gillin JC, Smith TL, Irwin M, Kripke DF, Schuckit M. EEG sleep studies in "pure" primary alcoholism during subacute withdrawal: relationships to normal controls, age, and other clinical variables. *Biol Psychiatry* 1990;**27**(5):477–88.
73. Brower KJ, Aldrich MS, Robinson EA, Zucker RA, Greden JF. Insomnia, self-medication, and relapse to alcoholism. *Am J Psychiatry* 2001;**158**(3):399–404.
74. Le Bon O, Verbanck P, Hoffmann G, Murphy JR, Staner L, De Groote D, et al. Sleep in detoxified alcoholics: impairment of most standard sleep parameters and increased risk for sleep apnea, but not for myoclonias—a controlled study. *J Stud Alcohol* 1997;**58**(1):30–6.
75. Williams HL, Rundell Jr. OH. Altered sleep physiology in chronic alcoholics: reversal with abstinence. *Alcohol Clin Exp Res* 1981;**5**(2):318–25.
76. Heinz A, Beck A, Mir J, Grusser SM, Grace AA, Wrase J. Alcohol craving and relapse prediction: imaging studies. In: Kuhn CM, Koob GF, editors. *Advances in the neuroscience of addiction*. Boca Raton, FL: Frontiers in Neuroscience; 2010.
77. Brower KJ. Insomnia, alcoholism and relapse. *Sleep Med Rev* 2003;**7**(6):523–39.
78. Gillin JC, Smith TL, Irwin M, Butters N, DeModena A, Schuckit M. Increased pressure for rapid eye movement sleep at time of hospital admission predicts relapse in nondepressed patients with primary alcoholism at 3-month follow-up. *Arch Gen Psychiatry* 1994;**51**(3):189–97.
79. Brower KJ, Aldrich MS, Hall JM. Polysomnographic and subjective sleep predictors of alcoholic relapse. *Alcohol Clin Exp Res* 1998;**22**(8):1864–71.
80. Kubota T, De A, Brown RA, Simasko SM, Krueger JM. Diurnal effects of acute and chronic administration of ethanol on sleep in rats. *Alcohol Clin Exp Res* 2002;**26**(8):1153–61.
81. Hill SY, Reyes RB. Effects of chronic and acute ethanol administration on sleep in laboratory rats. *J Stud Alcohol* 1978;**39**(1):47–55.
82. Mendelson WB, Hill SY. Effects of the acute administration of ethanol on the sleep of the rat: a dose-response study. *Pharmacol Biochem Behav* 1978;**8**(6):723–6.
83. Sharma R, Sahota P, Thakkar M. Rapid tolerance development to the NREM sleep promoting effect of alcohol. *Sleep* 2014;**37**(4):821–4.
84. Thakkar MM, Engemann SC, Sharma R, Sahota P. Role of wake-promoting basal forebrain and adenosinergic mechanisms in sleep-promoting effects of ethanol. *Alcohol Clin Exp Res* 2010;**34**(6):997–1005.

85. Ticho SR, Stojanovic M, Lekovic G, Radulovacki M. Effects of ethanol injection to the preoptic area on sleep and temperature in rats. *Alcohol* 1992;**9**(3):275–8.
86. Mukherjee S, Kazerooni M, Simasko SM. Dose-response study of chronic alcohol induced changes in sleep patterns in rats. *Brain Res* 2008;**1208**:120–7.
87. Mukherjee S, Simasko SM. Chronic alcohol treatment in rats alters sleep by fragmenting periods of vigilance cycling in the light period with extended wakenings. *Behav Brain Res* 2009;**198**(1):113–24.
88. Ehlers CL, Slawecki CJ. Effects of chronic ethanol exposure on sleep in rats. *Alcohol* 2000;**20**(2):173–9.
89. Rouhani S, Dall'Ava-Santucci J, Bajenaru O, Emmanouilidis E, Tran G, Manicom R, et al. Effects of muscimol or homotaurine on sleep-wake states in alcohol-dependent rats during withdrawal. *Pharmacol Biochem Behav* 1998;**59**(4):955–60.
90. Rouhani S, Emmanouilidis E, Tran G, Durlach J, Payan C, Fermanian J, et al. Circadian variations in vigilance states in the alcohol-dependent rat. *Physiol Behav* 1990;**48**(5):637–40.
91. Mendelson WB, Majchrowicz E, Mirmirani N, Dawson S, Gillin JC, Wyatt RJ. Sleep during chronic ethanol administration and withdrawal in rats. *J Stud Alcohol* 1978;**39** (7):1213–23.
92. Veatch LM. Disruptions in sleep time and sleep architecture in a mouse model of repeated ethanol withdrawal. *Alcohol Clin Exp Res* 2006;**30**(7):1214–22.
93. Gitlow SE, Bentkover SH, Dziedzic SW, Khazan N. Persistence of abnormal REM sleep response to ethanol as a result of previous ethanol ingestion. *Psychopharmacologia* 1973;**33**(2):135–40.
94. Hill SY, Reyes RB. Effects of L-tryptophan and ethanol on sleep parameters in the rat. *Psychopharmacology (Berl)* 1978;**58**(3):229–33.
95. Hunsley MS, Palmiter RD. Norepinephrine-deficient mice exhibit normal sleep-wake states but have shorter sleep latency after mild stress and low doses of amphetamine. *Sleep* 2003;**26**(5):521–6.
96. Prospero-Garcia O, Criado JR, Henriksen SJ. Pharmacology of ethanol and glutamate antagonists on rodent sleep: a comparative study. *Pharmacol Biochem Behav* 1994;**49** (2):413–16.
97. Breese GR, Criswell HE, Carta M, Dodson PD, Hanchar HJ, Khisti RT, et al. Basis of the gabamimetic profile of ethanol. *Alcohol Clin Exp Res* 2006;**30**(4):731–44.
98. Wallner M, Hanchar HJ, Olsen RW. Low dose acute alcohol effects on GABA A receptor subtypes. *Pharmacol Ther* 2006;**112**(2):513–28.
99. Aguayo LG, Peoples RW, Yeh HH, Yevenes GE. GABA(A) receptors as molecular sites of ethanol action. Direct or indirect actions? *Curr Top Med Chem* 2002;**2**(8):869–85.
100. Siggins GR, Roberto M, Nie Z. The tipsy terminal: presynaptic effects of ethanol. *Pharmacol Ther* 2005;**107**(1):80–98.
101. Winsky-Sommerer R, Vyazovskiy VV, Homanics GE, Tobler I. The EEG effects of THIP (Gaboxadol) on sleep and waking are mediated by the GABA(A)delta-subunit-containing receptors. *Eur J Neurosci* 2007;**25**(6):1893–9.
102. Radulovacki M, Virus RM, Djuricic-Nedelson M, Green RD. Adenosine analogs and sleep in rats. *J Pharmacol Exp Ther* 1984;**228**(2):268–74.
103. Nagy LE, Diamond I, Casso DJ, Franklin C, Gordon AS. Ethanol increases extracellular adenosine by inhibiting adenosine uptake via the nucleoside transporter. *J Biol Chem* 1990;**265**(4):1946–51.
104. Aeschbach D, Matthews JR, Postolache TT, Jackson MA, Giesen HA, Wehr TA. Dynamics of the human EEG during prolonged wakefulness: evidence for frequency-specific circadian and homeostatic influences. *Neurosci Lett* 1997;**239**(2-3):121–4.
105. Choi DS, Cascini MG, Mailliard W, Young H, Paredes P, McMahon T, et al. The type 1 equilibrative nucleoside transporter regulates ethanol intoxication and preference. *Nat Neurosci* 2004;**7**(8):855–61.

106. Dar MS. Brain adenosinergic modulation of acute ethanol-induced motor impairment. *Alcohol Alcohol Suppl* 1993;**2**:425–9.
107. Dar MS. Mouse cerebellar adenosinergic modulation of ethanol-induced motor incoordination: possible involvement of cAMP. *Brain Res* 1997;**749**(2):263–74.
108. Dar MS. Modulation of ethanol-induced motor incoordination by mouse striatal A(1) adenosinergic receptor. *Brain Res Bull* 2001;**55**(4):513–20.
109. Saeed DM. Co-modulation of acute ethanol-induced motor impairment by mouse cerebellar adenosinergic A1 and GABA(A) receptor systems. *Brain Res Bull* 2006;**71**(1–3):287–95.
110. Newton PM, Messing RO. Intracellular signaling pathways that regulate behavioral responses to ethanol. *Pharmacol Ther* 2006;**109**(1–2):227–37.
111. Dunwiddie TV, Masino SA. The role and regulation of adenosine in the central nervous system. *Annu Rev Neurosci* 2001;**24**:31–55.
112. Hack SP, Christie MJ. Adaptations in adenosine signaling in drug dependence: therapeutic implications. *Crit Rev Neurobiol* 2003;**15**(3–4):235–74.
113. Concas A, Mascia MP, Cuccheddu T, Floris S, Mostallino MC, Perra C, et al. Chronic ethanol intoxication enhances [3H]CCPA binding and does not reduce A1 adenosine receptor function in rat cerebellum. *Pharmacol Biochem Behav* 1996;**53**(2):249–55.
114. Prediger RD, da Silva GE, Batista LC, Bittencourt AL, Takahashi RN. Activation of adenosine A1 receptors reduces anxiety-like behavior during acute ethanol withdrawal (hangover) in mice. *Neuropsychopharmacology* 2006;**31**(10):2210–20.
115. Kaplan GB, Bharmal NH, Leite-Morris KA, Adams WR. Role of adenosine A1 and A2A receptors in the alcohol withdrawal syndrome. *Alcohol* 1999;**19**(2):157–62.
116. Batista LC, Prediger RD, Morato GS, Takahashi RN. Blockade of adenosine and dopamine receptors inhibits the development of rapid tolerance to ethanol in mice. *Psychopharmacology (Berl)* 2005;**181**(4):714–21.
117. Benington JH, Kodali SK, Heller HC. Stimulation of A1 adenosine receptors mimics the electroencephalographic effects of sleep deprivation. *Brain Res* 1995;**692**(1–2):79–85.
118. Landolt HP, Gillin JC. Sleep abnormalities during abstinence in alcohol-dependent patients. Aetiology and management. *CNS Drugs* 2001;**15**(5):413–25.
119. Basheer R, Strecker RE, Thakkar MM, McCarley RW. Adenosine and sleep-wake regulation. *Prog Neurobiol* 2004;**73**(6):379–96.
120. Franks HM, Hagedorn H, Hensley VR, Hensley WJ, Starmer GA. The effect of caffeine on human performance, alone and in combination with ethanol. *Psychopharmacologia* 1975;**45**(2):177–81.
121. Hasenfratz M, Bunge A, Dal PG, Battig K. Antagonistic effects of caffeine and alcohol on mental performance parameters. *Pharmacol Biochem Behav* 1993;**46**(2):463–5.
122. Liguori A, Robinson JH. Caffeine antagonism of alcohol-induced driving impairment. *Drug Alcohol Depend* 2001;**63**(2):123–9.
123. Fillmore MT. Alcohol tolerance in humans is enhanced by prior caffeine antagonism of alcohol-induced impairment. *Exp Clin Psychopharmacol* 2003;**11**(1):9–17.
124. Irwin M, Gillin JC, Dang J, Weissman J, Phillips E, Ehlers CL. Sleep deprivation as a probe of homeostatic sleep regulation in primary alcoholics. *Biol Psychiatry* 2002;**51**(8):632–41.
125. Erblich J, Earleywine M, Erblich B, Bovbjerg DH. Biphasic stimulant and sedative effects of ethanol: are children of alcoholics really different? *Addict Behav* 2003;**28**(6):1129–39.
126. Martin CS, Earleywine M, Musty RE, Perrine MW, Swift RM. Development and validation of the Biphasic Alcohol Effects Scale. *Alcohol Clin Exp Res* 1993;**17**(1):140–6.
127. Greiffenstein P, Mathis KW, Stouwe CV, Molina PE. Alcohol binge before trauma/hemorrhage impairs integrity of host defense mechanisms during recovery. *Alcohol Clin Exp Res* 2007;**31**(4):704–15.

128. Courtney KE, Polich J. Binge drinking in young adults: data, definitions, and determinants. *Psychol Bull* 2009;**135**(1):142–56.
129. Sprow GM, Thiele TE. The neurobiology of binge-like ethanol drinking: evidence from rodent models. *Physiol Behav* 2012;**106**(3):325–31.
130. Ritzmann RF, Tabakoff B. Dissociation of alcohol tolerance and dependence. *Nature* 1976;**263**(5576):418–20.
131. Snell LD, Szabo G, Tabakoff B, Hoffman PL. Gangliosides reduce the development of ethanol dependence without affecting ethanol tolerance. *J Pharmacol Exp Ther* 1996;**279**(1):128–36.
132. Anji A, Kumari M. Supplementing the liquid alcohol diet with chow enhances alcohol intake in C57 BL/6 mice. *Drug Alcohol Depend* 2008;**97**(1–2):86–93.
133. Sharma R, Engemann SC, Sahota P, Thakkar MM. Effects of ethanol on extracellular levels of adenosine in the basal forebrain: an in vivo microdialysis study in freely behaving rats. *Alcohol Clin Exp Res* 2010;**34**(5):813–18.
134. Sharma R, Sahota P, Thakkar MM. Lesions of the basal forebrain cholinergic neurons attenuates sleepiness and adenosine after alcohol consumption. *J Neurochem* 2017;**142**:710–20.
135. Van Dort CJ, Baghdoyan HA, Lydic R. Adenosine A(1) and A(2A) receptors in mouse prefrontal cortex modulate acetylcholine release and behavioral arousal. *J Neurosci* 2009;**29**(3):871–81.
136. Rhodes JS, Best K, Belknap JK, Finn DA, Crabbe JC. Evaluation of a simple model of ethanol drinking to intoxication in C57BL/6J mice. *Physiol Behav* 2005;**84**(1):53–63.
137. Kamdar NK, Miller SA, Syed YM, Bhayana R, Gupta T, Rhodes JS. Acute effects of naltrexone and GBR 12909 on ethanol drinking-in-the-dark in C57BL/6J mice. *Psychopharmacology (Berl)* 2007;**192**(2):207–17.
138. Linsenbardt DN, Boehm SL. Agonism of the endocannabinoid system modulates binge-like alcohol intake in male C57BL/6J mice: involvement of the posterior ventral tegmental area. *Neuroscience* 2009;**164**(2):424–34.
139. Mulligan MK, Rhodes JS, Crabbe JC, Mayfield RD, Adron HR, Ponomarev I. Molecular profiles of drinking alcohol to intoxication in C57BL/6J mice. *Alcohol Clin Exp Res* 2011;**35**(4):659–70.
140. Linsenbardt DN, Boehm SL. Role of novelty and ethanol history in locomotor stimulation induced by binge-like ethanol intake. *Alcohol Clin Exp Res* 2012;**36**:887–94.
141. Linsenbardt DN, Moore EM, Griffin KD, Gigante ED, Boehm SL. Tolerance to ethanol's ataxic effects and alterations in ethanol-induced locomotion following repeated binge-like ethanol intake using the DID model. *Alcohol Clin Exp Res* 2011;**35**(7):1246–55.
142. Melon LC, Boehm SL. GABAA receptors in the posterior, but not anterior, ventral tegmental area mediate Ro15-4513-induced attenuation of binge-like ethanol consumption in C57BL/6J female mice. *Behav Brain Res* 2011;**220**(1):230–7.
143. Moore EM, Boehm SL. Site-specific microinjection of baclofen into the anterior ventral tegmental area reduces binge-like ethanol intake in male C57BL/6J mice. *Behav Neurosci* 2009;**123**(3):555–63.
144. Sharma RS, Sahota P, Thakkar MM. A single episode of binge alcohol drinking causes sleep disturbance, disrupts sleep homeostasis and downregulates equilibrative nucleoside transporter 1. *J Neurochem* 2018;**146**:304–21.
145. Kalinchuk AV, McCarley RW, Stenberg D, Porkka-Heiskanen T, Basheer R. The role of cholinergic basal forebrain neurons in adenosine-mediated homeostatic control of sleep: lessons from 192 IgG-saporin lesions. *Neuroscience* 2008;**157**(1):238–53.
146. Landolt HP. Sleep homeostasis: a role for adenosine in humans? *Biochem Pharmacol* 2008;**75**(11):2070–9.

147. Kaur S, Junek A, Black MA, Semba K. Effects of ibotenate and 192IgG-saporin lesions of the nucleus basalis magnocellularis/substantia innominata on spontaneous sleep and wake states and on recovery sleep after sleep deprivation in rats. *J Neurosci* 2008;**28**(2):491–504.
148. Zeitzer JM, Nishino S, Mignot E. The neurobiology of hypocretins (orexins), narcolepsy and related therapeutic interventions. *Trends Pharmacol Sci* 2006;**27**(7):368–74.
149. Finelli LA, Baumann H, Borbely AA, Achermann P. Dual electroencephalogram markers of human sleep homeostasis: correlation between theta activity in waking and slow-wave activity in sleep. *Neuroscience* 2000;**101**(3):523–9.
150. Borbely AA, Achermann P. Sleep homeostasis and models of sleep regulation. *J Biol Rhythms* 1999;**14**(6):557–68.
151. Kalinchuk AV, Porkka-Heiskanen T, McCarley RW, Basheer R. Cholinergic neurons of the basal forebrain mediate biochemical and electrophysiological mechanisms underlying sleep homeostasis. *Eur J Neurosci* 2015;**41**(2):182–95.
152. Porkka-Heiskanen T, Strecker RE, Thakkar M, Bjorkum AA, Greene RW, McCarley RW. Adenosine: a mediator of the sleep-inducing effects of prolonged wakefulness. *Science* 1997;**276**(5316):1265–8.
153. Palchykova S, Winsky-Sommerer R, Shen HY, Boison D, Gerling A, Tobler I. Manipulation of adenosine kinase affects sleep regulation in mice. *J Neurosci* 2010;**30**(39):13157–65.
154. Sharma R, Engemann S, Sahota P, Thakkar MM. Role of adenosine and wake-promoting basal forebrain in insomnia and associated sleep disruptions caused by ethanol dependence. *J Neurochem* 2010;**115**(3):782–94.
155. Alanko L, Stenberg D, Porkka-Heiskanen T. Nitrobenzylthioinosine (NBMPR) binding and nucleoside transporter ENT1 mRNA expression after prolonged wakefulness and recovery sleep in the cortex and basal forebrain of rat. *J Sleep Res* 2003;**12**(4):299–304.
156. Faingold CL. The Majchrowicz binge alcohol protocol: an intubation technique to study alcohol dependence in rats. *Curr Protoc Neurosci* 2008. Chapter 9: Unit9.28.
157. Majchrowicz E. Induction of physical dependence upon ethanol and the associated behavioral changes in rats. *Psychopharmacologia* 1975;**43**(3):245–54.
158. Sharma R, Bradshaw K, Sahota P, Thakkar MM. Acute binge alcohol administration reverses sleep-wake cycle in Sprague Dawley rats. *Alcohol Clin Exp Res* 2014;**38**(7):1941–6.
159. Bertola A, Mathews S, Ki SH, Wang H, Gao B. Mouse model of chronic and binge ethanol feeding (the NIAAA model). *Nat Protoc* 2013;**8**(3):627–37.
160. Olive MF, Mehmert KK, Nannini MA, Camarini R, Messing RO, Hodge CW. Reduced ethanol withdrawal severity and altered withdrawal-induced c-fos expression in various brain regions of mice lacking protein kinase C-epsilon. *Neuroscience* 2001;**103**(1):171–9.
161. Urayama A, King K, Gaskin FS, Farr SA, Banks WA. Effects of chronic ethanol administration on brain interstitial fluid levels of Methionine-enkephalin as measured by microdialysis in vivo. *Peptides* 2006;**27**(9):2201–6.
162. Nordskog BK, Hammarback JA, Godwin DW. Diurnal gene expression patterns of T- type calcium channels and their modulation by ethanol. *Neuroscience* 2006;**141**(3):1365–73.
163. Pan W, Barron M, Hsuchou H, Tu H, Kastin AJ. Increased leptin permeation across the blood-brain barrier after chronic alcohol ingestion. *Neuropsychopharmacology* 2008;**33**(4):859–66.
164. Watson WP, Little HJ. Identification of distinct components, with different time courses, of the changes in response to convulsive stimuli during ethanol withdrawal. *J Pharmacol Exp Ther* 1995;**272**(2):876–84.

165. O'Neill SE, Parra GR, Sher KJ. Clinical relevance of heavy drinking during the college years: cross-sectional and prospective perspectives. *Psychol Addict Behav* 2001;**15**(4):350–9.
166. Jennison KM. The short-term effects and unintended long-term consequences of binge drinking in college: a 10-year follow-up study. *Am J Drug Alcohol Abuse* 2004; **30**(3):659–84.
167. Johnson BA. Medication treatment of different types of alcoholism. *Am J Psychiatry* 2010;**167**(6):630–9.
168. Armitage R, Hoffmann R, Conroy DA, Arnedt JT, Brower KJ. Effects of a 3-hour sleep delay on sleep homeostasis in alcohol dependent adults. *Sleep* 2012;**35**(2):273–8.
169. Brower KJ, Hoffmann R, Conroy DA, Arnedt JT, Armitage R. Sleep homeostasis in alcohol-dependent, depressed and healthy control men. *Eur Arch Psychiatry Clin Neurosci* 2011;**261**(8):559–66.
170. Krishnan HR, Sakharkar AJ, Teppen TL, Berkel TD, Pandey SC. The epigenetic landscape of alcoholism. *Int Rev Neurobiol* 2014;**115**:75–116.
171. Ponomarev I, Wang S, Zhang L, Harris RA, Mayfield RD. Gene coexpression networks in human brain identify epigenetic modifications in alcohol dependence. *J Neurosci* 2012;**32**(5):1884–97.
172. Wolstenholme JT, Warner JA, Capparuccini MI, Archer KJ, Shelton KL, Miles MF. Genomic analysis of individual differences in ethanol drinking: evidence for non-genetic factors in C57BL/6 mice. *PLoS One* 2011;**6**(6):e21100.
173. Wong CC, Mill J, Fernandes C. Drugs and addiction: an introduction to epigenetics. *Addiction* 2011;**106**(3):480–9.
174. Warnault V, Darcq E, Levine A, Barak S, Ron D. Chromatin remodeling—a novel strategy to control excessive alcohol drinking. *Transl Psychiatry* 2013;**3**:e231.
175. Paxinos G, Watson C. *The rat brain in stereotaxic coordinates*. 6th ed. Boston: Academic Press; 2007.
176. Pandey SC, Sakharkar AJ, Tang L, Zhang H. Potential role of adolescent alcohol exposure-induced amygdaloid histone modifications in anxiety and alcohol intake during adulthood. *Neurobiol Dis* 2015;**82**:607–19.
177. Sakharkar AJ, Zhang H, Tang L, Shi G, Pandey SC. Histone deacetylases (HDAC)-induced histone modifications in the amygdala: a role in rapid tolerance to the anxiolytic effects of ethanol. *Alcohol Clin Exp Res* 2012;**36**(1):61–71.
178. You C, Zhang H, Sakharkar AJ, Teppen T, Pandey SC. Reversal of deficits in dendritic spines, BDNF and Arc expression in the amygdala during alcohol dependence by HDAC inhibitor treatment. *Int J Neuropsychopharmacol* 2014;**17**(2):313–22.

Index

Note: Page numbers followed by "*f*" and "*t*" refer to figures and tables, respectively.

R